W9-BZC-917

ESD Program Management

A Realistic Approach to Continuous Measurable Improvement in Static Control

G. Theodore Dangelmayer
AT&T Network Systems

To all of those people in the electronics industry who have yet to fully understand the value of a comprehensive Electrostatic Discharge (ESD) control program and to those who are striving for continuous improvement.

Library of Congress Catalog Card Number 90-34522
ISBN 0-442-23794-4

Printed in the United States of America

Van Nostrand Reinhold
115 Fifth Avenue
New York, New York 10003

Van Nostrand Reinhold International Company Limited
11 New Fetter Lane
London EC4P 4EE, England

Van Nostrand Reinhold
102 Dodds Street
South Melbourne, Victoria 3205, Australia

Nelson Canada
1120 Birchmount Road
Scarborough, Ontario M1K 5G4, Canada

16 15 14 13 12 11 10 9 8 7 6 5 4 3 2

Library of Congress Cataloging-in-Publication Data

Dangelmayer, Ted.
ESD program management: a realistic approach to continuous measurable improvement in static control/Ted Danglemayer.
p. cm.
Includes bibliographical references.
ISBN 0-442-23794-4
1. Electronic apparatus and appliances—Protection. 2. Electric discharges.
3. Manufacturing processes—Management. I. Title.
TK7870.D27 1990
621.381—dc20 90-34522
CIP

Contents

Foreword

In today's electronics business, managing an ESD program is an integral part of a complete quality program. In fact, any electronics firm without an active ESD program puts itself and its customers at risk. This book illustrates one good example of the detail and dedication to quality that AT&T expects within its own operations and from its suppliers.

Writing of the book began at a time when Ted Dangelmayer was burdened with many demands. These demands were from AT&T's own operations, internal suppliers, external suppliers, customers and others looking for a better understanding of the phenomenon of ESD, its impact and, most of all, ways to control and manage it. In a way, this book is a response to these demands by making available a reader friendly document that distills the hard-won experiences of Ted and AT&T.

The information and methods in this book have been gained at no small cost and produce results that far exceed expenses. There is, however, a caveat: Success will not be obtained unless there is real management commitment. This means management must allocate the necessary resources and provide active support to ensure that training, auditing, reporting, tracking and an aggressive corrective action program all take place successfully.

Ted is an internationally recognized authority, and you will benefit greatly by listening to his advice and following his recommendations.

Leonard J. Winn
Manager, Quality Control
AT&T Merrimack Valley Works
North Andover, Massachussetts

Preface

Electrostatic discharge (ESD) events can have serious detrimental effects on the manufacture and performance of microelectronic devices, the systems that contain them, and the manufacturing facilities used to produce them. Submicron device technologies, high system operating speeds, and factory automation are making ESD control programs a critical factor in the quality and reliability of ESD-sensitive products.

The detrimental effects that ESD has on sensitive electronic devices and assemblies are well documented and now receive considerable attention, both in design and in handling procedures. Furthermore, AT&T, the Department of Defense, and many informed companies place stringent design, handling, and packaging requirements on suppliers. However, to date, insufficient emphasis has been placed on properly managing a comprehensive program to mitigate the effects of ESD, and yet, the success of an ESD program depends on how well it is managed.

Successful ESD program management requires a total system approach, one that is interwoven into every aspect of the manufacturing process from product design to customer acceptance. This book presents a comprehensive proven plan that is suitable for any electronics manufacturing company, large or small. It was developed over a twelve year period at AT&T.

It has been written mainly for engineers and managers in the electronics manufacturing industry who are contemplating the organization of an ESD control program or who are interested in strengthening an existing program. It will also be beneficial to a wide variety of occupational disciplines including trainers, maintenance technicians, installation and service personnel, packaging engineers, test engineers, process engineers, product engineers, design engineers, employees who assemble and test sensitive products, quality control personnel, purchasing agents, and anyone involved with automation.

This book should be read from cover to cover and then used as a reference document. Many of the concepts are interwoven and, therefore, best understood by reading the entire book. The "Points To Remember," found at the end of each chapter, have been added to help the reader use the book as a reference and to highlight some of the more important aspects of each chapter.

Chapter 1 offers an overview of the steps to be taken to implement an ESD control program. Sixteen steps are described, beginning with the necessity to understand the technology and ending with advice to continuously improve the program.

The next two chapters, 2 and 3, provide the reader with the foundation of knowledge for building an ESD control program. Chapter 2 contains a description of the critical factors in managing the program, while Chapter 3 has a tutorial on the basic principles of electrostatics.

Chapter 4 describes the scientific manufacturing experiments performed at AT&T to learn about the effects of ESD on devices and how control procedures can be implemented to solve the problem. This chapter shows how to collect data for justifying an ESD control program, as well as the results from our experiments. Chapter 14 parallels Chapter 4. It contains payback data and a description of the benefits derived from the successful implementation of an ESD control program.

Chapters 5 through 11 describe the different parts of the ESD control program. They also represent the ESD coordinator's tools for implementing a total system solution. These chapters include information and advice on how to implement each part of the program and how to manage the parts as an integrated program. The coordinator's tools consist of material on designed-in protection, a test lab, a handbook containing realistic requirements, an auditing program, purchasing, and training.

Chapter 12 (packaging considerations) and Chapter 13 (automation) contain examples of how an ESD coordinator uses the program management tools to solve problems.

At the end of the book is a glossary of terms, a reference listing and an index.

Acknowledgments

A very special thanks goes to Leonard Winn who literally made the writing of this book possible. He not only provided the necessary resources, but also gave me the encouragement that was needed at the outset.

Special recognition is due to those who contributed significantly to the writing of this book. Terry Welsher is responsible for writing Chapters 3, 5, and 12. Chapter 6 was written largely by Dick Morrow with help from Dawn Dube and Tom Diep.

Thanks go to the editors, and especially Don Ford, whose editorial advice was a valued resource throughout the writing of the book. He not only helped organize the book, but also rewrote a number of chapters. Victor Goodman helped with the editing and rewriting in the final stages. Marge Stanbury was the copy editor who kept us on track.

Sincere appreciation is extended to all of the members of management who have continually supported this project over the years. It could not have succeeded without their strong commitment. I am especially grateful for the professional freedom given to me to manage the project with few restrictions.

Many colleagues have contributed significantly, either directly or indirectly, to the knowledge base of this book and deserve acknowledgement. Thanks go to Burt Unger, now with Bellcore, for the technical expertise he shared and the research he and his group performed. Thanks also go to Ed Jesby for his help in developing the auditing and training techniques, and to Joe Doucette for his work as coauthor of the first ESD control handbook which was later the basis of the AT&T *Electrostatic Discharge Control Handbook*. Special acknowledgement goes to George Abate, Lew Parham, and Ed Campiglio for their sense of pride and the knowledge we gained in their relentless surveillance of the factory as Plant ESD Inspectors.

Thanks go to Jim McKinney, Jim Brown, Pat Bennett, and Elizabeth McMillian from the AT&T Document Development Organization for their production support.

Thanks go to Bob Newton for the vast majority of the photography and to Roger Culliford for the photographs in Chapter 12.

My most sincere and heartfelt appreciation goes to my family for their support and forbearance of the countless hours of solitude. Adding a book to an already overflowing schedule resulted in many sacrifices. My wife, Lindy, has always been an invaluable source of strength and now an editor, therapist, and advisor. I am especially grateful for the understanding my sons Andrew and Peter displayed. My only regret is that my father passed away during the writing of the book and did not see it completed.

Introduction

I have been asked repeatedly how I became so heavily involved with Electrostatic Discharge (ESD) control and how we established such a strong commitment at AT&T. These questions are best answered by briefly reviewing the evolution that took place over many years.

My first involvement was in 1978 when we discovered extensive damage to thin film resistors. Initially the source of the damage was unknown but later was found to be caused by ESD during the cleaning operation in mass soldering machines.

At the time, I was on loan to AT&T Bell Laboratories from AT&T Network Systems in a failure mode analysis capacity and had previously been a manufacturing test engineer. As a result of being in AT&T Bell Laboratories, I had the resources readily available to conduct a thorough investigation. This proved to be the beginning of a fascinating and fruitful journey.

Initially I thought it would be a six-month assignment and that all I would need to do would be to install a few wrist straps and tabletops and then move on. This is a common misconception of many people, even today. That was eleven years ago, and I have been working on ESD related matters on a full-time basis since 1982. The more we learn the more involved the assignment becomes, requiring added time and effort.

With the resistor experience in mind, obvious questions arose regarding the extent of ESD damage to silicon devices. I decided to answer the questions scientifically by conducting a battery of designed experiments. These were done on the manufacturing floor under normal conditions. Simultaneously, we initiated a number of pilot programs in manufacturing areas to measure the effect that ESD could have on yields.

These studies, which are detailed in Chapter 4, continued for approximately four years, and as more data became available, management commitment grew. The initial commitment to conduct the studies came from the manager of thin film engineering. Based on preliminary information and instinct, he allocated the necessary resources.

Widespread support of middle management emerged as a result of the irrefutable evidence gained from these early manufacturing experiments and because of the research conducted by AT&T Bell Laboratories. Management learned that yields improved and operating costs declined, with a return on investment of over 1000 percent. They also learned that ESD damage could result in reliability failures that would affect customers.

The knowledge gained in these early years would later prove to be the basis of our realistic handling requirements as well as the auditing program that verifies compliance. The experiments told us with certainty that, if the procedures we used are adhered to, product yields will improve accordingly. Therefore, it was possible to develop an auditing program based on the goal of ensuring 100 percent compliance with procedures and know that it would be cost-effective.

In 1982 we sought a commitment from the top levels of management in the North Andover plant to implement a systematic control plan throughout the facility. We knew this would require a substantial investment of resources at a time when budgets were tight, but we were confident that the overwhelming evidence would be sufficient justification.

The presentation took two hours because of the probing and pertinent questions. We were well prepared because of the detailed work of the previous four years and the experience gained. Consequently, the Manufacturing Vice President and his staff gave us full support. In fact, it was at this point that I was assigned to manage the project on a full-time basis.

This commitment is a reflection of the type of management we have enjoyed throughout the project and has been one of the critical factors in our success.

In 1983 it became clear that a corporate thrust would be necessary to gain the full benefits of systematic control and to avoid redundant effort. So we put together a corporate symposium in Denver to stimulate interest and to disseminate information. This ultimately led to the formation of a Corporate ESD Committee by Corporate Engineering and the beginning of a companywide effort. Manufacturing and design areas from approximately 50 different AT&T locations were represented on the committee. We were able to create and publish requirements with which all locations could agree to adhere. The corporate committee has proven to be the primary reason for the unified approach to ESD control in the company and continues to be an invaluable resource.

The committee was gaining momentum in 1984 when divestiture took place. At that time, we suffered a significant setback when the AT&T Bell Laboratories ESD Studies Group was divested to support the Regional Bell Telephone Operating Companies. This meant that we no longer had a research and development resource and that our customers, the Bell Operating Companies, now had one of the best in the world.

In 1985, this adversity became an opportunity to establish an even stronger management commitment at higher levels in the company. We knew we needed help from upper management and were debating the right course of action when the Manufacturing Vice President who had helped us earlier came forward and asked how we were doing in regard to designed-in protection. When he heard about the transfer and the possible long-term implications, he immediately took charge of the situation and asked me to plan on giving a presentation to his manager in two weeks. This led to giving the same presentation at another level higher, and one month later at the staff meeting of the Executive Vice President. It was decided at that meeting to put me on the Bell Laboratories agenda for their staff meeting and on the the Quality Council agenda. The Quality Council is a group of Vice Presidents that represent all business entities within the company. The Bell Laboratories executive staff is responsible for all design related matters and resources.

During the Bell Laboratories staff meeting, it was decided that the ESD Studies Group should be reestablished. Subsequently, Bell Laboratories personnel were assigned to define a charter for the group and to formally submit a request for funding. It was approved shortly thereafter and fully staffed by the end of the year. Since then, the group has emerged as an invaluable resource within AT&T, and by 1987 had become recognized as one of the industry leaders in the field with a number of technical contributions.

In summary, this evolution of management commitment from the grass roots to very high levels has resulted in a broad base of support essential to long-term success. Without this support, the roadblocks that developed along the way would have been insurmountable. The support is based on demonstrable quality enhancements during production, the growing need for designed-in protection, and customer satisfaction.

In the reports to management on quality improvements, we proved that ESD events can damage even robust devices and that an investment in ESD control during manufacturing would lower the overall operating costs. AT&T would be able to produce more reliable products at a lower cost and develop another competitive edge in the process.

The program was shown to reduce defect levels by more than 50 percent on many product lines, while serving as an insurance program against unanticipated production problems. We have faced situations with ultrasensitive devices where failure thresholds were 20 volts and the yields approached zero. In these cases, any company would find it impossible to meet production planning levels and risk losing the sale of that product line.

However, with the expertise and resources of an ESD control program in place, we found solutions quickly and were able to meet production schedules. Of course, with over ten years of experience, we can now anticipate and prevent many of the problems that might arise.

In addition, we used the data, and our growing knowledge of electrostatics, to keep the program realistic and, thereby, the costs down. We avoided purchasing unnecessary control equipment by carefully examining how devices fail and how to prevent it. These cost-effective measures helped sustain management commitment.

As our customers began to realize how ESD control provided them with better products, the high visibility of the program became an unexpected benefit. On tours of the plant, customers saw the consistent use of protective items, such as wrist straps, throughout the operation. This highly visible aspect of an ESD control program is viewed by customers, and correctly so, as a reflection of our strong commitment to quality and reliability.

When management learned that ESD control enhances the company image, customer satisfaction became another major driving force behind our program. In fact, now that many customers understand the reliability and cost implications associated with ESD control, they place stringent design and handling requirements on our work. This is because more reliable products permit our customers to reduce their costs and improve their service position with fewer repairs.

Thus, we at AT&T found that an ESD control program makes good business sense. The program produces a substantial return on investment while strengthening business agreements with customers who relate ESD control to reliability. The program has proven to be a worthwhile investment and good insurance against high failure rates.

This book is about the ESD control program that emerged from this evolutionary process. Most of the pioneering work was done at the AT&T Network Systems plant in North Andover, Massachusetts (also called the Merrimack Valley Works). Looking back, this manufacturing facility was an ideal test location for evaluating, understanding, and developing an ESD control program. This is because of the diverse range of products produced at this facility and because of its size.

It is a large facility with approximately 2 million square feet of floor space and 8000 employees. It is a telecommunications transmission equipment factory, and the range of products produced or handled are shown in Figures 1 and 2. We receive silicon wafers from other locations, produce hybrid integrated circuits, fabricate bare printed wiring board (PWB) assemblies, produce PWB assemblies that are heavily populated with integrated circuits, and produce sophisticated transmission equipment for central offices. Extensive testing and process verification are done at all stages of manufacturing to ensure that high quality and reliability standards are being met.

Some readers are certain to say, "But my plant is different from AT&T's Merrimack Valley Plant. Will their procedures and ESD control program work at my plant?" Our evidence says that the procedures will work very well in other manufacturing facilities.

The program that was developed at the Merrimack Valley plant was adopted by the corporate ESD committee and successfully implemented at a number of AT&T plants.

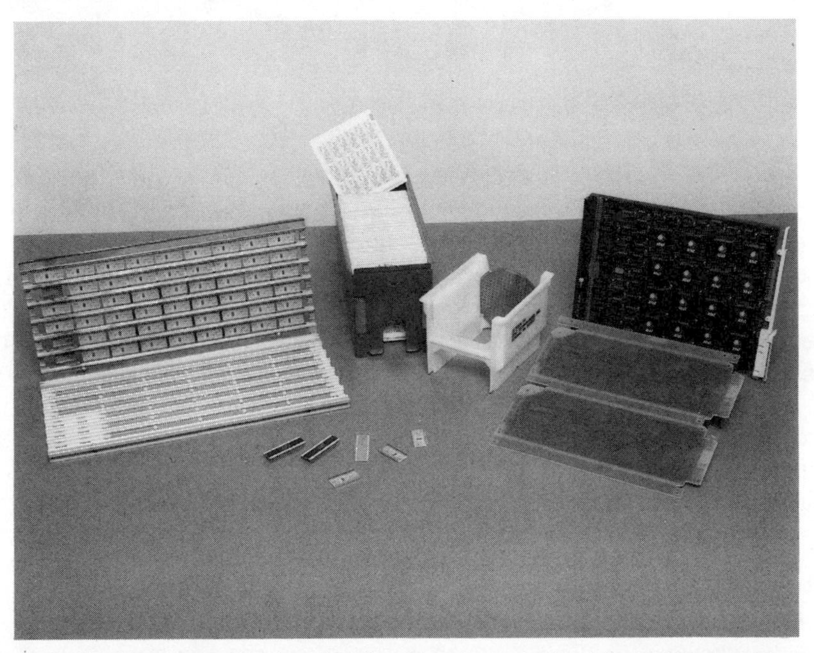

Figure 1. Products at the Merrimack Valley Plant from right to left: PWB assembly, bare PWB, wafer, hybrid integrated circuit (HIC) substrates in a handling magazine, HICs, and HICs in encapsulation trays.

Figure 2. A completed digital transmission equipment system

For instance, both the Reading Works in Pennsylvania and the Denver Works in Colorado experienced the same continuous improvement after adopting the auditing procedures described in Chapters 8 and 9. Compare Figures 3, 4 and 5 and note the remarkable similarity. The experience gained at Merrimack Valley made it possible for Denver and Reading to not only duplicate the continual improvement but to do it in less time.

Although the ESD control procedures in this book have proven to be successful in different plants, implementing and using them requires a commitment to quality control and a total system approach that is of paramount importance as emphasized in Chapter 1.

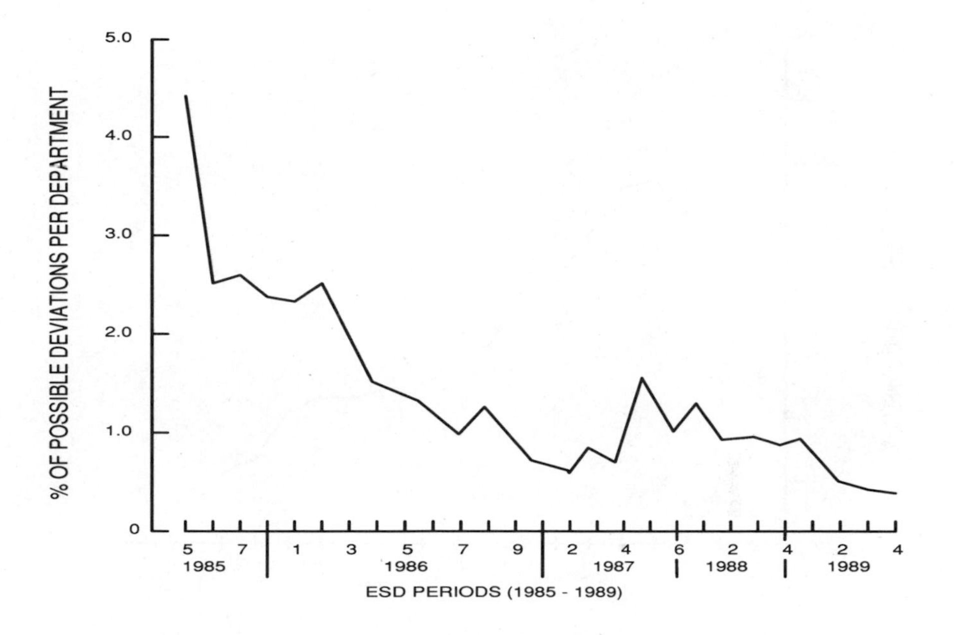

Figure 3. Merrimack Valley plant ESD trend analysis chart (see Chapter 9 for details)

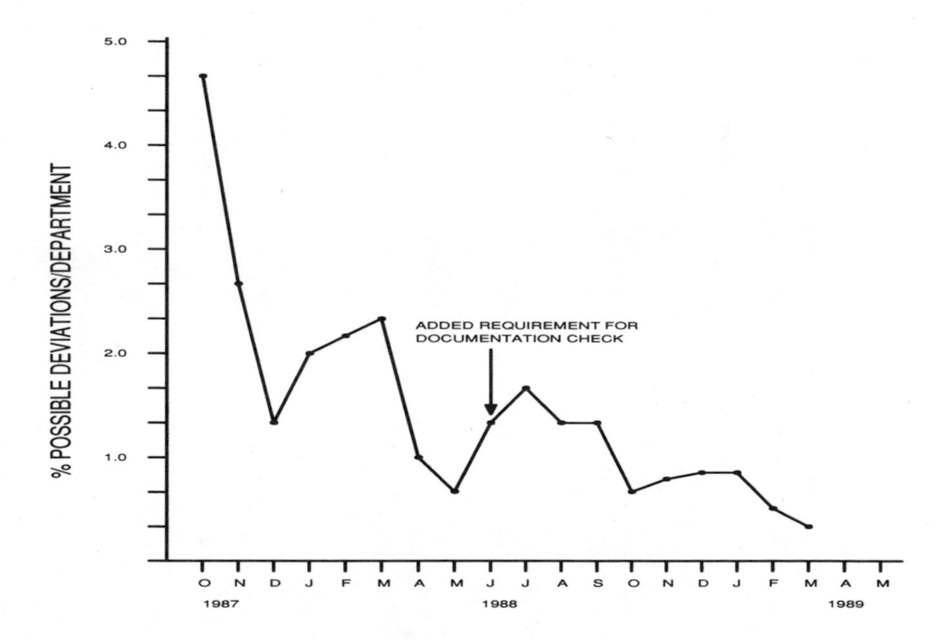

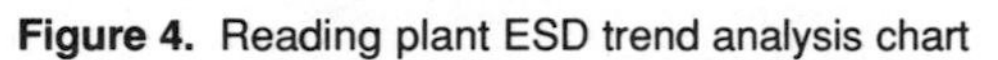
Figure 4. Reading plant ESD trend analysis chart

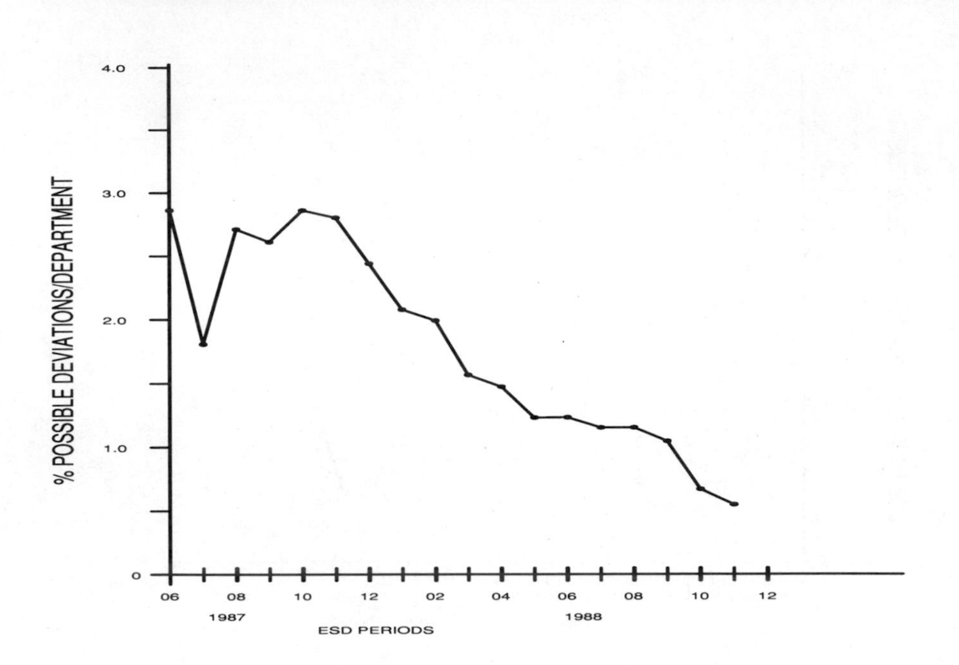

Figure 5. Denver plant ESD trend analysis chart

Chapter 1

Twelve Critical Factors in ESD Program Management

Developing, implementing, and managing a successful ESD program requires a total system approach that extends from product design to customer acceptance. The program will need to be well managed and woven into every aspect of the manufacturing process in order to produce lasting success. In fact, ***a well-managed program can be far more effective than one well stocked with expensive supplies***. The twelve critical factors described in this chapter (Table 1-1) form the basis of successful ESD program management.

The first four factors listed in this chapter give the project its much needed organization and authority during the start-up and implementation phase, as well as afterwards during the long-term continual-improvement phase. These factors consist of a written implementation plan, management commitment, a full-time coordinator, and an active ESD committee. The next five factors described are the coordinator's essential tools: realistic requirements, a training program, an auditing program, a test lab, and an extensive communication effort. The three remaining factors are management principles that will help the program run more efficiently. These are systemic planning, human factors engineering, and continuous improvement.

The ESD coordinator must be aware of the twelve critical factors and fully understand their significance. ***The factors need to be managed by the coordinator just as actively as the controls to protect sensitive devices and assemblies***. Sound management of these factors will produce a cost-effective program and sustained success.

Think of this chapter as an introduction to the twelve critical factors. By highlighting them in advance and out of context, they can be carried in the back of one's mind somewhat like a mental checklist. While some factors will stand out more than others because they are given separate chapters, all are very important. Each one is woven into the fabric of the ESD control program, so each will be found in various contexts throughout the book. They are very important ideas, central to successful ESD program management. It can be very useful to return to this chapter repeatedly while studying the book, implementing a program, or evaluating the status of an ongoing program.

Table 1-1. The Twelve Critical Factors for Successful ESD Program Management

Factor One	An Effective Implementation Plan
Factor Two	Management Commitment
Factor Three	A Full-Time Coordinator
Factor Four	An Active Committee
Factor Five	Realistic Requirements
Factor Six	Training for Measurable Goals
Factor Seven	Auditing Using Scientific Measures
Factor Eight	ESD Test Facilities
Factor Nine	A Communication Program
Factor Ten	Systemic Planning
Factor Eleven	Human Factors Engineering
Factor Twelve	Continuous Improvement

The Twelve Critical Factors

Factor One — An Effective Implementation Plan

The success of an ESD program depends on how well it is implemented. The best of programs can fail in the absence of a sound implementation plan. Therefore, it is critically important to develop an effective implementation plan in writing. Begin by developing a thorough understanding of the concepts in this book. Be sure that the outline of the implementation plan reflects the intent of the chapters that follow. The details of the plan should then be written down in the form of an action plan, to document individual responsibilities, deadlines, and progress. When finished, it will organize the massive undertaking of implementing a new or examining an ongoing program into a series of smaller projects, as well as give a first approximation of the work schedule. Be sure to include suppliers, distributors, and subcontractors when developing the plan because they are extensions of the program and must comply with its particulars. In fact, these companies should be chosen based on verifiable compliance with proper ESD procedures and approved packaging materials.

The plan is built around the other eleven factors described in this chapter and the steps described in Chapter 2. As the reader studies this book, talks with members of the ESD committee, and surveys the manufacturing plant, details to the plan should be inserted under the appropriate category such as management commitment, a test lab, purchasing, automation, and so on. With each entry, the coordinator should include thoughts on who will work on each task and when.

By organizing the implementation effort in this way, the coordinator can see the larger picture and be less apt to get mired in details. Priorities can be set more effectively, tasks can be delegated to appropriate coworkers, and the timing of events can be regulated with an eye to how well employees are progressing with the programs.

Factor Two — Management Commitment

ESD control transcends the entire company, its suppliers, and subcontractors. Therefore, it is critically important to have support from all levels of management, especially from the top levels. In this way, a coordinated effort can be established swiftly and efficiently to implement the details of the plan. Otherwise, the numerous roadblocks that can develop along the way will become insurmountable and the program will fail.

Consequently, ***management commitment must be actively sought*** and then periodically reaffirmed for the program to succeed in the long run. In a large plant, there might be two or three hundred managers working in different capacities in the organizational hierarchy. A small plant, however, might have only two or three managers. In either case, managers at all levels who have authority over employees and have commitment to the ESD solution are a major part of the program's success. Without a clear and strong commitment early on from top management, there can be no long–term effort at solving the problem. Even if a few middle- or lower-level managers drag their feet at first, they will join the effort fairly quickly if they see that top management believes strongly in the program.

Once the commitment has been established, it should become common knowledge to all employees. It should be visible in the form of a signed statement describing the company's program to prevent ESD damage to devices and the nature of the ESD solution. It helps if management reissues this statement every year. Also, evidence of management commitment is seen when there is a budget to implement items such as training, purchasing, auditing, and so on. Included in this budget must be the money for a full-time coordinator.

Factor Three — A Full-Time Coordinator

Successfully implementing the ESD program requires a full-time effort by a well-qualified professional. This is a critical element, especially in the early stages. Later on, a part-time effort may be sufficient in smaller companies. However, in large companies the task cannot be done effectively on a part-time basis. Studying the technology, selecting and purchasing the needed equipment, preparing the procedures handbook and manuals, building a training program, and putting the critical factors into place are no trivial task. The problem is compounded by the fact that very few engineers understand the technology, and even fewer understand the risks. The ESD coordinator must serve as a consultant to all of the engineering disciplines in addition to overseeing the plan. Implementing an effective ESD management program requires a dedicated effort to reap the financial benefits.

The ESD coordinator should be a member of the quality organization. In this way, the program becomes a global responsibility transcending all manufacturing and engineering organizations and touching the entire manufacturing work flow from design to the finished product. Figure 1-1 illustrates a typical manufacturing flowchart that includes ESD considerations and extends from design to customer acceptance.

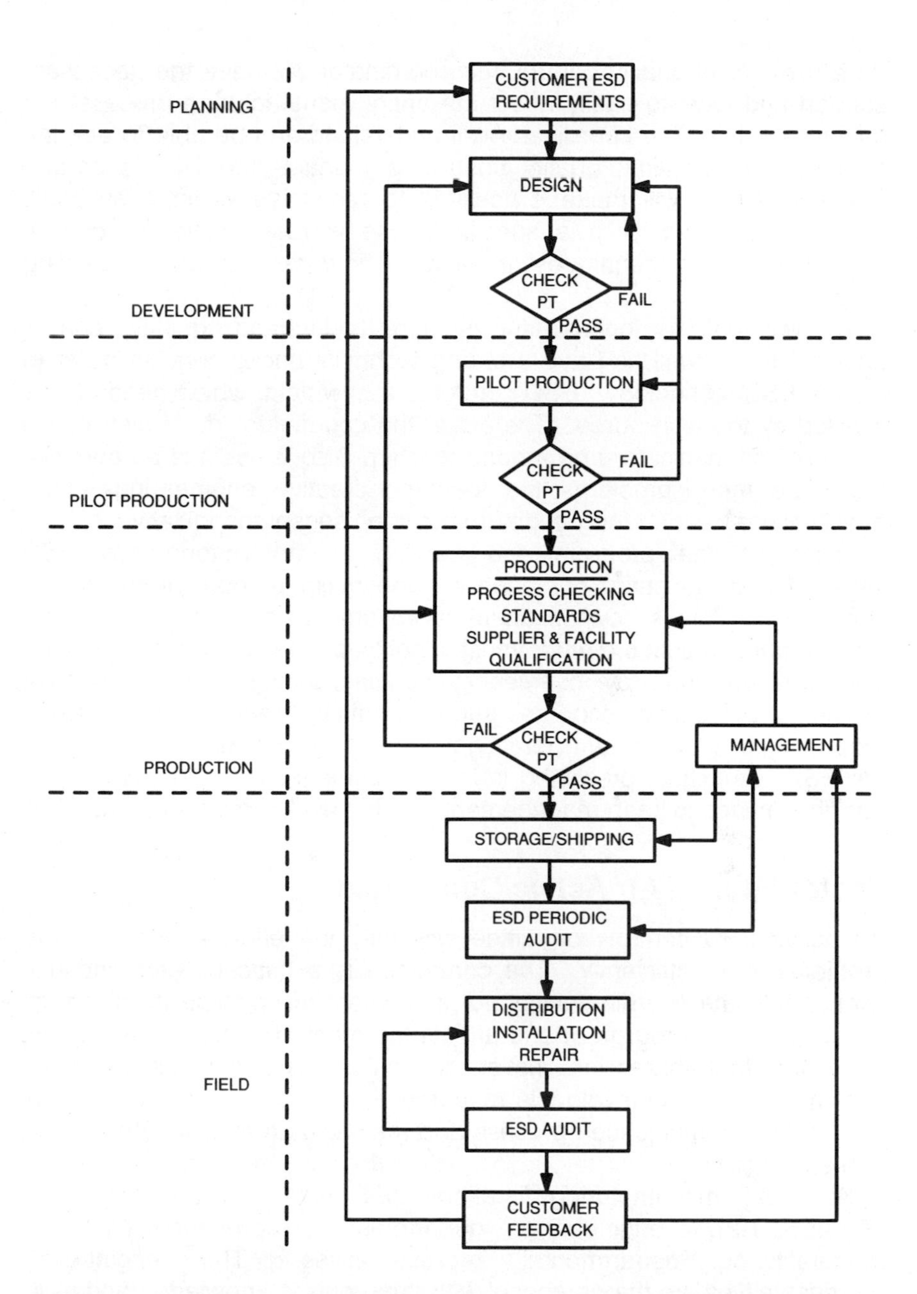

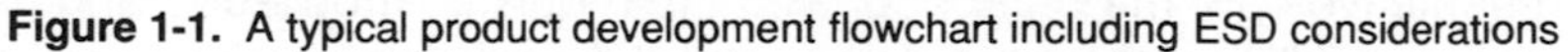

Figure 1-1. A typical product development flowchart including ESD considerations

As a member of quality control, the coordinator will have the necessary support and tools to easily affect the entire manufacturing process and business entity. The coordinator will understand and be able to use the concepts of statistical quality control to manage the ESD program. These concepts will make it possible to recognize when a systemic change is needed or if a specific cause is responsible for current problems. These techniques are also an integral part of the auditing program.

A well-qualified coordinator should understand quality control engineering as well as have a strong technical background in order to master ESD technology. ESD control is a new field, which needs to be learned by the work force. Therefore, the coordinator must also be an effective communicator and a good teacher. Also, since it is a new field, there are many problems that demand creative engineering. Other qualifications needed for the job are common sense and initiative.

Managing the program and serving as an authoritative ESD consultant to the entire company are the major responsibilities of the coordinator. This is how continual improvement happens. To manage the program, the coordinator must identify problems and solve them. This can be done by overseeing auditing, using data to pinpoint problems, delegating resources such as training, reporting and making proposals to management, directing purchasing, directing the efforts of the ESD committee, preparing training documentation, and using every possible means to keep management and the work force informed.

Factor Four — An Active Committee

An active ESD control committee will unify the effort and help solve problems more efficiently. The committee is a critical factor because even a full-time coordinator cannot implement or upgrade the program alone; the task is too great and affects too many disciplines. A working committee that shares information and enlists help from many experts and managers is an invaluable resource for the coordinator. It is a vital part of the communication process and results in an appropriate sharing of responsibility.

At AT&T manufacturing locations, we have two subcommittees (Figure 1-2): engineering departmental representatives and manufacturing departmental representatives. The engineering representatives are the personnel with the requisite knowledge and skills for solving ESD problems. They include packaging engineers, test engineers, an ergonomic expert, trainers, safety representatives, product engineers, and process engineers. The manufacturing representatives

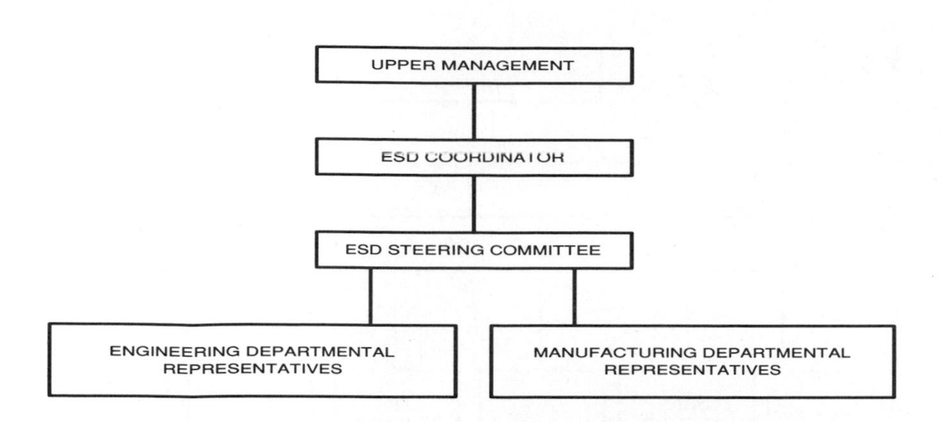

Figure 1-2. Local ESD control committee organizational flowchart

consist primarily of shop supervisors who coordinate the ESD effort in their department.

In general, the committee is responsible for developing policies and procedures. These responsibilities include promoting ESD awareness and control, ensuring local compliance with ESD instructions, identifying and resolving ESD issues, and maintaining a team of local ESD specialists to assist in solving related problems.

Large corporations with more than one manufacturing and assembly plant should also have a corporate committee to unify the company's efforts across plant boundaries. Figure 1-3 shows how this committee is staffed and organized at AT&T. It includes both manufacturing and research and development (R&D). Led by cochairpersons, subgroups include communications, control procedures and documentation, standards, consultants, and R&D.

The corporate committee is an effective means of sharing solutions to common problems and minimizing redundant effort. This approach provides a constancy of purpose that makes it possible to develop consensus standards and establish a unified approach to the needs of the corporation.

Factor Five — Realistic Requirements

The ESD control requirements must be realistic and formally documented, for they are the foundation of the entire plan (see Chapter 7, "Realistic Requirements"). All activities, procedures, and support documents are based on these requirements. It is, therefore, critically important for the requirements to be well documented in easy-to-understand language and mindful of training needs. Consistent

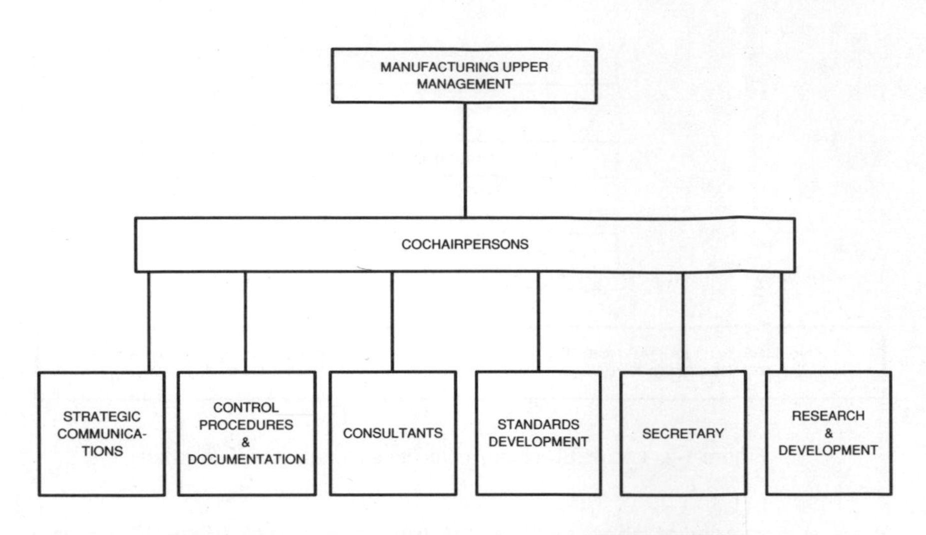

Figure 1-3. Corporate ESD control organizational flowchart

compliance with proper procedures depends on a complete understanding of the requirements; thus, thorough documentation is essential. Furthermore, the requirements must be realistic to be enforceable.

Written as a handbook, the requirements can also serve as a text during training and as a common reference for all employees. The information in the handbook is then a comprehensive statement of the ESD control program. Training is based on it, the auditing checklist is written to ensure compliance with it, and the employee's work is structured from it.

Although the requirements are extensive and complete, they must also be realistic so that people can follow them easily. They must be ***written in such a way that human error is improbable***. This is accomplished by explaining procedures clearly and by placing a high priority on human factors engineering (Factor Eleven) whenever possible.

In addition, support manuals should be written for activities such as process checking, auditing, and maintaining facilities. These documents should also be comprehensive and realistic.

Factor Six — Training for Measurable Goals

Training is an obviously critical factor in successful ESD control and must be a primary consideration at all times. A training program built on measurable goals derived from the auditing program allows the coordinator to aggressively pursue the identification and resolution of training needs. The auditing results clearly identify when training is needed, who needs training, what needs to be taught, and whether the training was successful.

Special emphasis should be placed on training engineers, maintenance technicians, and first-line supervisors. Given their involvement and highly visible positions, these people must set an example for others to follow. Instead, they are often the worst violators of proper ESD procedures and undermine the programs in an unconscious manner. A sound training program can correct this situation.

During the development and scheduling of training courses, it is vitally important to take into account the Three Principles of the Psychology of Training and Learning.[1] Using them as a guide will greatly enhance the effectiveness of the training program. These principles are discussed in more detail in Chapter 11 and are listed below:

Principle One: Train students only in order to affect a measurable change in work behavior.

Principle Two: Motivate students to improve learning.

Principle Three: Take into consideration the fact that students tend to forget information and skills not used regularly.

Factor Seven — Auditing Using Scientific Measures

Auditing is the binding force behind a sound program and is critical to a program's long-term success. Its mere presence spurs compliance and a strong management commitment that fosters continuous improvement. Published reports can motivate managers and engineers towards improving the program in their department.

The reports provide the coordinator with the tools necessary to effectively manage and maintain the program. They make it possible to easily identify problems and then solve them. Chapter 8, "Implementing an Auditing Program," and Chapter 9, "Using Auditing Results to Manage the ESD Program," provide examples of how major problems can be identified and permanently resolved.

The selection of the auditor (see Chapter 8) is critical. This person must be able to withstand peer pressure and to report all deviations as initially detected. The objective is to protect sensitive products from ESD damage by supplying management with valid information that can be used for swift corrective action.

Using statistical sampling techniques, auditing measures departmental compliance with the prescribed ESD control procedures. These are the procedures from the handbook and the manuals that employees are trained to follow. The statistical unit of measure for the program is the deviation from prescribed procedure.

The procedures are transformed into a questionnaire type of checklist. In addition to this checklist, the auditing program consists of an auditing inspector, a manual, a portable test cart, and software for filing and organizing data. The collected data is printed in graph form as either a trend chart or a Pareto analysis. The coordinator uses these graphs to spot trends, identify and pinpoint problems, and report progress to management. As stated above, these reports are also an invaluable training tool. The net result of an auditing program is continual improvement.

Factor Eight — ESD Test Facilities

Having adequate testing capability is a critical tool for the coordinator. It allows the coordinator to use electrical tests to scientifically evaluate many aspects of the program and its success. For instance, testing is an integral part of such activities as auditing, qualifying equipment and sensitive components for purchase, defining effective requirements and procedures, inspecting incoming control products, solving manufacturing problems, providing failure mode analysis on devices and systems, demonstrating during training, and testing and qualifying devices or systems prior to shipment.

While some testing can be very sophisticated, much of the testing recommended in this book is basic. This is in keeping with the program's philosophy of being realistic. For instance, by testing and qualifying ESD control equipment, greater standardization of the auditing test procedures can be achieved. In fact, the program at AT&T is set up so that all of the periodic requalification tests done by manufacturing process checkers can be accomplished with a wrist strap tester. This one idea lowers maintenance costs considerably, makes the test easier to perform, and simplifies training. (See Chapter 10 for more details.)

The test facilities on hand will depend on one's budget, plant size, and testing goals. Chapter 6 describes a variety of test equipment and one method of setting up three types of testing facilities, such as a field audit kit that fits in a suitcase, a general lab that includes a portable test cart, and an analytic/failure mode analysis (FMA) laboratory.

Factor Nine — A Communication Program

Effective communication is a vital element in successful ESD control and is ***one of the most challenging critical factors***. Coordinators often underestimate the difficulty or fail to recognize the importance of establishing a communication program. A sound communication program must be developed at the outset and actively managed at all times. For instance, a quality auditor once asked why a certain requirement had changed three times in nine months. In reality, the requirement had been published three years earlier and had never been changed; it was merely the auditor's understanding that changed. This is typical of the immense difficulties associated with effective communication.

Therefore, the coordinator should take advantage of every available channel of communication, take advantage of every opportunity to keep people aware of the ESD problem, demonstrate that progress is being made towards its solution, and post quality improvement charts and graphs where all can view them. It should be assumed that people need to know. What might seem obvious to a coordinator, such as how to test a bench top or adjust a wrist strap, is not necessarily obvious to others.

Examples of communication possibilities include publicizing the fact that a department had zero deviations or sending an ESD bulletin when a new type of wrist strap is to be used. Information about the ESD control program should be included in the introductory materials for new employees. Publish a policy statement annually. Invite the local press to visit a class or a training demonstration. Give reports frequently to management, both in writing and verbally, during business results meetings. Coordinators should be able to find many additional ways of communicating with all involved parties.

Most plants have all of the necessary channels of communication available. These include signs, posters, bulletins, video displays, a public address system, classrooms and meeting rooms, capabilities for publishing handbooks and manuals, supervisors, quality process checkers or their equivalent, a newsletter, interoffice mail for memos and letters, and electronic mail. Less common and very effective is a Quality Fair (See Chapter 11, "Training for Measurable Goals.")

Factor Ten — Systemic Planning

The diverse elements of an ESD control program form a total system that will ultimately determine the success of the program. It is critically important to realize that ***each element is part of an integrated whole*** rather than separate distinct entities. A change in any part of the program will have a ripple effect on other elements. Conversely, there are times when the program must be changed to effect desired improvement. Therefore, systemic planning becomes yet another aspect of the program that requires constant awareness and management on the part of the coordinator.

First, by planning ahead for the ripple effect of a change, the coordinator can anticipate its total consequence. This will lessen confusion and prevent the creation of new problems while trying to solve an existing problem. For example, consider the implication of a relatively minor change in wrist straps in which you go from using three different sizes to using one adjustable wrist strap. That single change should prompt the following questions. Will people need additional training in how to adjust the new strap? Must the section on wrist straps in the handbook be rewritten? Does this change affect the inspector's manual or checklist? What is the most effective way to communicate this change? Can the new wrist strap be tested in the same manner as the ones being replaced? Must new test equipment be purchased? If a new test is warranted, what written documents must be changed? Will one adjustable wrist band really fit everybody? Will employees find the wrist straps comfortable and safe? Should there be backup wrist straps for very small or very large people? Can the new wrist strap be purchased without changing our long-term relationship with the current wrist strap vendor?

Secondly, the ripple effect in an integrated program often masks the root cause of a problem. Not finding the root cause in a program will mean facing the same problem again and again.

For example, auditing uncovered a problem with heelstraps. Further study revealed that many employees were wearing them incorrectly. The obvious solution would have been additional training. However, due to the complicated nature of the heelstrap, the training problem would have been endless. The root cause of the problem was really in the design of the heelstrap and not in the training methods. After the heelstrap was replaced with a simplified version, human error became improbable and the problem was permanently solved. See Chapter 9, "Using Auditing Results to Manage the ESD Program," for additional examples.

Factor Eleven — Human Factors Engineering

The employees' ability to comply with the ESD control procedures is a major part of the ESD solution. Every aspect of the program that affects people must be engineered in such a way that all reasonable employee needs and desires are taken care of and that human error is improbable. For example, if the equipment is uncomfortable or inconvenient, employees will be less apt to comply with the procedures. In fact, failure to consider their needs could cause a catastrophic breakdown in the program. Considering their needs through human factors engineering is also a critical factor, one which must be examined in all aspects of the program.

Asking different pilot groups of employees to trial-test all equipment and procedures is a primary technique of human factors engineering. This helps the coordinator understand employees' needs and builds a body of knowledge on effective solutions. For example, trial testing revealed that some types of wrist strap materials produced a rash on some employees. This discovery avoided a serious problem.

Our experience with trial-testing has proven to us that employees want to do their jobs well, which includes complying with procedures. Giving employees a say on matters concerning convenience and safety, whether it be a choice of color or the best length for wrist strap cords, costs little and will reap great rewards.

Factor Twelve — Continuous Improvement

Continuous improvement of each of the previous eleven critical factors is an essential part of a sound ESD control program. Implemented effectively, the critical factors will produce a cost–effective program. However, it is the continuous improvement of those factors that will sustain the success. Many companies, failing to recognize the importance of this, have undertaken control programs with enthusiasm only to let them deteriorate into a state of disrepair and total ineffectiveness. Consequently, the funds expended for the program have been wasted and none of the quality improvements have been realized. Often the deficiencies go undetected because there is no auditing program in effect.

In contrast, the ultimate goal of ESD control and all other manufacturing quality control efforts must be satisfied customers through better products, services, and costs. This is not a fixed goal like winning a road race. Continual improvement is an endless process of meeting one goal after another.

Fixed goals are an important first approximation when working toward continual improvement. For the first 5 years, the primary goal should be to achieve zero deviations from prescribed ESD control procedures. This measurable goal allows for the setting of priorities, putting the coordinator in charge of tackling the most serious problem first, the next serious problem second, and so on. It also provides a straightforward way to report results. Chapter 9, "Using Auditing Results to Manage the ESD Program," describes this technique in a detailed way.

But what happens when there are zero deviations in the program? Is the project a success? Is the project completed? No! The project will have accomplished a monumental achievement when zero deviations are achieved and sustained. The project should not, however, be considered completed. There is always room for further improvement. No matter how good we are today, we can and must be even better tomorrow. Furthermore, ***failure to continuously improve the process will translate into complacency and deterioration***.

There are always better and more cost-effective techniques, new control products to evaluate, new solutions to consider, and better training techniques to incorporate. Add to this the trend toward devices of ever increasing sensitivity to ESD damage. Staying abreast of the technology becomes vital. Later chapters illustrate the extreme technical difficulties that these ultrasensitive devices present as well as the need to be prepared for them.

Points To Remember

- The twelve critical factors form the basis for successful ESD program management.

- The twelve critical factors should be kept in mind when studying this book, implementing the program, and managing the ongoing program.

- The twelve critical factors should be used as categories when writing the implementation plan. The plan should be written as an action plan.

- Management commitment should be obtained at all levels and periodically reenforced.

- The nature and massive scope of the project requires a full-time coordinator who is well qualified for the job.

- A local ESD control committee composed of both engineering specialists and manufacturing managers will enlist help from others and unify the effort.

- A written handbook consisting of realistic requirements is the backbone of the control program.

- A training program based on measurable goals teaches the awareness and skills necessary for employees to comply with procedures. The training program is based on the handbook.

- An auditing program using scientific measures is a binding force behind the whole program and fosters continual improvement.

- Adequate ESD test facilities make it possible for the coordinator to scientifically develop, manage, and direct the control program.

- An active communication program, maintained at all times, uses all available channels of communication to explain and publicize ESD control.

- Systemic planning is important. An integrated program such as ESD control can experience a ripple effect, where changes in one part of the program can ripple through other parts of the program, causing new problems or masking the root cause of a problem.

- Employees will comply with procedures more willingly when human factors engineering is incorporated. The goal is to engineer solutions in such a way that human error is improbable.

- A primary goal of the ESD control program is continuous improvement through continual attention to the twelve critical factors. An important interim goal is zero deviations. The ultimate goal must be satisfied customers.

Chapter 2

Implementing an ESD Control Program: The Basic Steps

The 16 basic steps introduced in this chapter will provide the basis for the design and implementation of effective ESD program management. Presenting the steps first in this chapter in a brief outline offers a conceptual overview before focusing on individual steps in later chapters. This is especially beneficial for manufacturing companies contemplating the organization of such a program or for those attempting to strengthen their commitment to controlling ESD.

Following these 16 steps in the order in which they are presented will help foster two of the key elements of any successful companywide program: commitment and communication. (See Chapter 1 for more details.) The 16 steps are designed to secure, from the start, a top-down commitment to the program. Later steps will involve middle and lower levels of management and, finally, all members of the work force. As each department in the organization enlists in the ESD program, a strong sense of program ownership will emerge. The result should be a measurable improvement in performance, not only in that department, but throughout the entire organization.

Good communication is vital to the smooth implementation of an effective program. Communication ensures that top management remains involved in the program's development and deployment and allows all responsible personnel, throughout the organization, to follow the program's progress. Ongoing communication enables everyone to share in a successful implementation and to contribute to future improvements in the program. In addition, continued communication, over time, enables individual operating departments to revise their own goals, to learn from the progress in other departments, and to share in the discovery of new technologies and new testing procedures. In short, communication helps everyone responsible for ESD control to participate in the continuous improvement of the organization's program.

The Basic Steps

Step 1: Study and Understand the Technology of ESD Control

When strengthening an existing program or before an effective program of ESD control can be developed, it is absolutely essential for the ESD coordinator to acquire a thorough understanding of the physics and engineering involved in the control of ESD. This will permit the establishment of effective requirements, realistic goals, and a program of control specifically designed for each facility.

A thorough understanding of the ESD technology will make it possible to develop cost-effective solutions for specific handling requirements. In addition, this understanding assures that solutions will be technically sound and will provide the required level of protection without incurring an unreasonable level of expense. In short, this knowledge will help avoid what might otherwise be overprotective insurance.

Developing a worthwhile training program, one that will contribute to the ESD control program's continued success, also requires a thorough understanding of the technology of the problem and its possible solutions. The training itself is valid, and valuable, only if it is based on sound technology. If employees can see through a weak technological argument, then they will question every step the program demands that they take.

It is especially important that the facility's ESD coordinator share this thorough understanding of ESD technology. The coordinator must be perceived as a credible authority on all aspects of the problem and its solution. Frequently, the coordinator will be required to answer a wide range of technical questions regarding the program. The ability to respond knowledgeably will help the coordinator more readily win acceptance among all employees and management. In turn, employee acceptance of the coordinator as a knowledgeable authority will contribute to smoother program implementation. This is especially true where ESD damage often occurs with no discernible sensation.

The first step in achieving the necessary level of understanding is to review the existing literature.[2-21] The proceedings of symposia sponsored by the EOS/ESD Association (an international organization) are a good place to begin. They include papers presented by companies such as AT&T Bell Laboratories, specific AT&T manufacturing locations, Bellcore, British Telecommunications, 3M, Honeywell, Texas Instruments, and other leaders in the field of ESD control technology.

A second step to understanding ESD technology might be to attend the EOS/ESD Symposium. EOS/ESD symposia provide one of the best forums for the presentation and discussion of timely information on the state of ESD control technology. Attending tutorials is also helpful in learning about advances in the control of ESD. There are many tutorials from which to choose, including those offered at the EOS/ESD Symposia.

These symposia and forums also offer access to professional ESD consultants who may be available to provide assistance in developing or upgrading a customized control program. It is particularly important that any consultant hired to develop such a program have substantial experience in implementation.

Every company planning to implement an ESD control program or strengthening an existing program should join the EOS/ESD Association. As a member, the ESD coordinator will be able to find help in identifying qualified consultants. Attendance at EOS/ESD events will lead to familiarity with other leading authorities in the field whose expertise may prove helpful at a later date. Some locations in the United States now have local chapter meetings of the EOS/ESD Association, another valuable opportunity to acquire ESD knowledge at little cost.

Step 2: Gather Scientific Evidence of the Economic Value of ESD Control

One of the most important steps in gaining or strengthening management commitment for the development or upgrading of the ESD program is the establishment of the program's short- and long-term economic benefits. Gathering the scientific evidence to support these benefits will also yield an even deeper understanding of the technology. Additional knowledge can be gained through participation in scientific experiments designed to demonstrate the benefits of ESD control. The experiments should include pilot studies in designated areas so that yield improvements can be documented. Experiments of this kind are discussed more fully in Chapter 4, "An Economic Analysis."

There is a wealth of literature available that focuses on the economic benefits of ESD control programs. For example, the EOS/ESD Association has recently published *An ESD Management Focus*,[22] a collection of selected papers from the first decade of EOS/ESD Symposia. These papers describe control programs, their economic value, and the training programs associated with them.

The information gathered in this way will allow for a preliminary estimate of the savings to be generated by your facility's ESD program. This estimate may be updated as additional information becomes available. Publishing preliminary findings and issuing subsequent updates may help solidify needed management support for the program.

Step 3: Establish an Active ESD Committee

Assuring the success of an ESD control program is work for more than one person. In order for the program to succeed, the ESD coordinator will need the assistance of an active committee.

This committee should include representatives from all of the key functions within the facility such as training, safety, engineering, operations, manufacturing, packaging, purchasing, and personnel.

The committee's task should be to help plan and implement each of the steps outlined in this chapter, and described more fully in this book. Of course, these steps are only a guide to ESD program development or expansion. The steps adopted by the committee should be tailored to fit the specific conditions of each manufacturing facility.

Once the committee has been established, it is important that specific responsibilities be assigned to each member. ***The coordinator cannot plan and implement a complete program alone***.

At first, the committee should meet as frequently as once a week. Reports of each week's meetings should be distributed to key management people responsible for implementation. Frequent reports of the committee's progress will help prepare the ground work necessary to win management's long-term commitment to the program.

Additional details of committee functions and responsibilities were presented in Chapter 1, "Twelve Critical Factors in ESD Program Management."

Step 4: Develop a General Implementation Plan

Once the ESD committee is in place, the next step is the development of a comprehensive implementation plan. Special emphasis should be placed on gathering and organizing the information necessary to secure top management support. Because ***any one aspect of the plan will affect all others***, it is essential that an exhaustive program for ESD control be developed at once.

One of the first steps in plan development is the identification of ESD sensitive areas and the engineers who share responsibility for those areas. As a second step, an action plan (including goals and an implementation schedule) can be prepared for each area simultaneously.

Manufacturing areas must be carefully surveyed and their ESD control implementation needs identified in thorough detail. This survey will provide a basis for estimating the actual implementation costs and for projecting the rate of return on investment to be delivered by the ESD control program. This information will be necessary to help establish top management commitment and the funding necessary for implementation.

Step 5: Write or Adopt a Set of Realistic Handling Requirements

It is vitally important to formally document and distribute the handling requirements for each manufacturing facility at the onset of implementation. These requirements need to be based on a thorough understanding of ESD technology, as described in Steps 1 and 2. They must take into account the level of training required for each manufacturing operation. The requirements must also reflect an understanding of human nature and must realistically address employee attitudes.

Handling requirements may vary with device sensitivity. Yet, it is frequently impractical to train all employees in the many different ways that devices of differing sensitivity must be handled. Instead, consideration should be given to organizing work areas in the plant based upon device sensitivity, such that employee training will be consistent throughout each work area. Each area should be equipped to protect the most sensitive device that will be handled there. Install the equipment and train all employees in an area to use the equipment installed in that area. Thus, the equipment provided will introduce added protection for the more sensitive devices, and the degree of training will be the same in each area. This kind of engineering solution will maximize the value of employee training. Additional information on training requirements can be found in Chapter 11, "Training for Measurable Goals."

Document all handling requirements before attempting to gain a broad management commitment. Once that commitment has been given, implementation will move rapidly and the handling requirements documentation will be required for consistent compliance with ESD control procedures. The documentation will also provide a valuable platform for good communication within the work force. Distribute the handling requirements to all employees immediately after gaining management commitment for the project.

Step 6: Prepare a Detailed Statement of Policy Including Individual Responsibilities

The company's corporate ESD policy statement should clearly stipulate the intention to control ESD in all operating environments. The policy should also ***include a clear definition of responsibility*** for every individual in the enterprise, from top management to all employees. See Appendix 1, "AT&T ESD Policy" for an example.

To ensure that the statement will be read by the widest possible audience, make certain that the entire statement fits on a single typewritten page. Plan to post the statement in appropriate and prominent places throughout the company and to distribute it, individually, to management personnel.

Be sure to gain management's complete commitment to the policy prior to publication.

Step 7: Prepare a Presentation for Management

Management reports must be concise, direct, and to the point, or they will be ineffective and the opportunity to win top-down support may be lost. The information should be current and should be well organized. Plan on finalizing the presentation after a meeting with management has been scheduled. Until then, facts should be updated continuously so that a formal report can be quickly prepared for distribution.

Information to be presented in the report includes the economic analysis used to estimate the potential savings engendered by ESD control in Step 2. The results of any experiments that may have been conducted at the facility should be included. Use these figures to estimate the return on investment through increased yields due to reduced static (ESD) induced failures during manufacturing.

The results of the literature search should be paraphrased and presented in a concise format so that management may observe trends in technology and implementation. This information should also be presented in such a way as to demonstrate the relevancy of the scientific literature to the facility for which the new or expanded ESD control program is intended.

The report should also include a reminder of the technological trend toward devices of ever increasing sensitivity. Even if no major problems are apparent in the facility today, as devices become more sensitive, problems are sure to occur.

A discussion of the intangible benefits that the program promises to deliver should also be included. They can be of great importance. As discussed in Chapter 14, "Payback and Benefits," one important intangible benefit of ESD control is a significant improvement in customer satisfaction.

The presentation should also include a brief synopsis of the proposed implementation plan and a systematic approach to the problem. Other obviously essential elements of the presentation are a proposed policy statement and a copy of the handling requirements.

Ask that management become involved at once and conclude with a clearly stated request for the desired commitment. That commitment is, after all, essential to the program's success. The report should recommend that the coordinator be appointed on a full-time basis to launch the program successfully. It can also be mentioned that in later years, depending on the size of the facility or company, it may not be necessary to maintain this as a full-time position. This appointment of a full-time coordinator is the first test of management's commitment to the success of the ESD control program.

Step 8: Establish a Top Management Commitment

As soon as a thorough, systematic, well coordinated program and presentation have been developed, the next essential detail is to request an opportunity to present the case as soon as it is practical to do so to top management. Prompt action is critical to the long-term success of the program.

When a program that involves the entire company is instituted, such as an ESD control program, the drive to implement and succeed must come from the top of the organization. The program is too important to the success of the company to be left to lower management. Effective ESD control requires the cooperation of every department, without regard to traditional areas of responsibility and "turf." In addition, the

ESD coordinator must be able to cross conventional lines of report and responsibility with top management's freely granted blessings. Therefore, the initial management presentation should take place as early as possible in the development of the program in order to clear any roadblocks that may inadvertently arise along the way.

The importance of gaining the early and active support of top management cannot be overemphasized. Make the presentation concise and to the point. The developers must include a direct request for approval of the new policy and procedures.

Step 9: Develop the Details of ESD Program Management

Once top management commitment has been established, it is time to develop the management details of a new or expanded ESD program. Because damage caused by ESD is both insidious and often undiscernible when it occurs, it is necessary to develop a coordinated control plan for purchasing, training, manufacturing, and shipping. In fact, a systematic approach to ***ESD control must extend from the initial design of the product through customer acceptance***. To achieve this virtual blanket of protection, the plan for all areas must be fully integrated.

This comprehensive approach requires that the actual conditions present within the organization and its suppliers be considered before the onset of implementation. Among the necessary steps that should be taken are the following:

- Select suppliers of sensitive devices and assemblies based on their ESD control program, the design of their products, and their protective packaging for shipment. Test methods for qualifying product design are described in Chapter 5, "Designed-In Protection and Product Testing." A supplier's control program should meet or exceed the standards set in this book. See Chapter 9 for an example of how to audit subcontractors or suppliers.

- Purchase ESD control equipment based on a thorough understanding of ESD control requirements. A purchasing plan should be instituted that is fully coordinated with all other aspects of the program (see Chapter 10, "Purchasing Guidelines: Finding the Hidden Costs and Problems"). Equipment specifications should always include training considerations and documentation implications.

- Establish and maintain, on-site, a test laboratory. The lab is required for the regular analysis of all items purchased, as well as for a thorough understanding of handling requirements and new ESD control technology. (See Chapter 6, "ESD Test Facilities.")

- Develop a customized training program that takes into account how people in the operation actually do their jobs. Ongoing training, for all levels of the work force, is essential in order to instill safe and proper work habits. (See Chapter 11, "Training for Measurable Goals.")

- Develop a quality control plan to ensure day-to-day compliance. It is vitally important that this plan include statistical sampling in the manufacturing area. In that way, the program will allow for timely verification that the intended procedures are being followed. (See Chapter 8, "Implementing an Auditing Program," and Chapter 9, "Using Auditing Results to Manage the ESD Program.")

Above all, for any plan to succeed, there must be a facilitywide commitment to strive for continuous improvement.

Step 10: Present the Program to Middle and Lower Management

The next phase in implementation is to expand the presentation of the program goals and practices to include middle and lower management. Reaching out to these management levels is the first step in developing broad, cross-departmental support for the program.

The middle management presentation is an opportunity to explain the details of the program to those who are going to be most directly responsible for the program's success. This presentation is the first forum in which to define and explain individual responsibilities. At the same time, the presentation offers a good opportunity to emphasize the need for a total team effort.

The ESD coordinator cannot successfully manage a comprehensive control program alone. He/she will need to ***delegate many of the responsibilities*** to different individuals. Clearly the success of that delegation requires selecting people whose primary responsibilities assure their involvement with ESD control.

For example, if engineering is responsible for providing the machinery for any given operation, then the same engineers should be responsible for providing the ESD control functions for that operation. If manufacturing is responsible for providing its own facilities, then that department should also be responsible for purchasing and maintaining the ESD control facilities used within that manufacturing environment. This is a highly effective way to delegate responsibility and to foster department by department ownership of a corporatewide program.

ESD is a major part of everyone's responsibility, much like safety. Delegating specific responsibilities is both necessary and reasonable, but requires management support. Details of the coordinator's responsibilities were discussed in Chapter 1, "Twelve Critical Factors in ESD Program Management."

Step 11: Begin Implementation of the Control Facilities

Once the details of the program have been laid out, management has been informed, and all participating departments have been assigned their ESD control responsibilities, the primary activity of the ESD control committee then becomes coordinating the implementation of control facilities such as wrist straps, grounded workstations, and dissipative tote trays.

The action plan that was developed earlier (see Step 4) can now be used to expedite implementation, by identifying critical items and making sure that they are implemented in a timely fashion. Following the action plan closely will help ensure that all implementation activities are coordinated for optimal efficiency.

Prepare and distribute status reports to the appropriate management levels on a regular basis. This will serve to keep them up-to-date on facility implementation throughout the corporation. Periodic reporting will also help to maintain the necessary emphasis on the project.

It is absolutely essential that ***savings realized at each stage of implementing the control procedures be fully documented as they are achieved***. This step is too often overlooked, even though clearly documented savings will maintain vital management interest in the project. As the savings promised by the program's earlier projections are actually realized, support among department heads will grow. At the same time, management's commitment, established when the project began, will find a continued basis for reinforcement.

Now is also the right time to begin awareness training, as described in Chapter 9. Awareness training should focus on both individual employees and on management supervisors. In addition, awareness training should include a program of ongoing communications that maintains interest in and concern for ESD control throughout both plants and corporate offices.

As part of that timely communications effort, a copy of top management's signed ESD policy statement should be distributed throughout the organization. Reissue the statement annually, to ensure that everyone continues to understand the importance to the operation of controlling ESD.

Step 12: Begin Training of the Entire Work Force, Including Management

Once facilities are in place, actual hands-on training of the work force can begin. The most effective method used is to train in stages. Proper organization of training materials and lessons is the key to acceptance, retention, and active participation.

Provide employees with a thorough explanation of why ESD control is necessary. Describe how each specific control program helps to prevent damage from ESD. Be especially clear in pointing out how each individual can contribute to the success of the total ESD program. It is essential that all questions be answered and that any fears, however irrational, be laid to rest.

A key element in training is to be certain that employees understand how the systems and equipment work. Employees must know how to operate new systems and equipment before they are challenged by their use in actual daily operations. For example, the necessity of personnel grounding and the technology behind wrist straps and conductive footwear should be explained. Provide ample reassurance of the safety of these devices, and explain how safety resistors work to protect an employee. Employees are often surprised to find that these devices are actually safer than a direct touch to electrical ground procedures where there is no current limiting resistor present. To stimulate and maintain interest further, include lively demonstrations whenever possible.

To achieve the highest level of positive response from the trainees and to sustain the desired change in behavior, complete training before asking employees to use ESD control devices. At this stage in the program's implementation, the requirement is to reach as many people as possible, as quickly as possible. Therefore, mass training techniques are not only in order, they are recommended. Many of these training techniques are described in detail in Chapter 11, "Training for Measurable Goals."

Step 13: Begin Using Facilities Immediately After Training

With the proper ESD control facilities in place and training complete, employees will be able to use these facilities as soon as they return to their jobs. Any delay in the application of what they have learned can seriously impair the training program's effectiveness and reduce employee acceptance.

It is particularly important not to allow employees to persist in improper or inadequate performance, even if a shortage of ESD control facilities should develop. By permitting inappropriate behavior for any reason, management risks appearing to condone that behavior. Unfortunately, once a way of performing a task has been learned, unlearning it is extremely difficult. For example, if employees are allowed to touch ground when a wrist strap has not been installed or are permitted to hold PWB assemblies by their faceplates because wrist straps are unavailable, potentially damaging habits will result and will prove hard to eliminate even when the proper equipment is in place.

During this early stage in the implementation phase, ***anticipate the need for on-the-job training***. Virtually every individual in the building will need some additional training, either from a management supervisor or from a professional trainer.

Consistency from the start is the key to training the work force. Always insist that the ESD program's documented requirements are followed to the letter. If employees are consistently unable to meet that standard of performance, consider whether the requirements might be unrealistic and need to be revised.

Step 14: Begin Formal Auditing of All Manufacturing Departments

After a 6-month grace period and a number of courtesy surveys have been completed, it is time to begin formal auditing of the progress in all departments.

Until all ESD control facilities have been installed and the basic work force training has been conducted, the ESD coordinator and several well trained engineers on the committee should conduct informal surveys of each operating site. In the course of these surveys, they should identify problem areas and classify them. Trouble spots can be grouped by the type of handling violation, with specific reference to either improperly trained employees or unacceptable facility installation.

A plan of corrective action for each area should be established, and the results entered as soon as they are available. This information should be shared informally with all involved departments, as well as with the representatives from engineering and manufacturing.

Meetings should be held to discuss whatever has been discovered in the course of the surveys. Open communication will help to gain the support of department personnel and will help them to develop a better understanding of the requirements. Then, when actual auditing begins, employees in each department will be better positioned to satisfy those requirements for handling procedures.

The experience gained during these surveys will also form the basis for more detailed auditing programs. These auditing programs will need to be enacted as the entire operation begins to follow the requirements of the ESD control program. Formal auditing can begin after the 6-month grace period. By then, most departments will have a thorough understanding of the program's requirements and procedures.

Step 15: Report Auditing Results, Graphically and With Regularity, to All Levels of Management

To ensure improvement, it is necessary to audit results carefully and consistently. Publication of the results will ensure that improvement is rapid as well.

Unfortunately, corporate decision makers often do not have time to read long reports; yet it is absolutely essential that management be kept up to date on the progress of the ESD control program throughout the operation. Continued top-down enthusiasm is needed if the program is to succeed over the long term.

Therefore, reports need to be short and to the point. Information must be presented graphically in order to highlight results such as comparisons of work force behavior and departmental performance. Graphic representations of this kind of information can instantly demonstrate the state of ESD control within the corporation.

Reports that are easy to read and convey information at a glance will inspire management to react to situations quickly and more effectively. ***Distribute these reports to all levels of management*** in order to maintain essential emphasis on the program goals and performance. Only if both top and middle management are kept informed of necessary adjustments to the program will appropriate corrective action be taken in a timely fashion. To that end, reports should be issued monthly or with each audit of every department.

Graphs showing quality improvement can also be displayed where the direct labor work force can see them. This will aid in enlisting universal support for the ESD program.

Step 16: Continuously Improve the Process

New, improved, and more cost-effective solutions to the problems of ESD related damage are continuously being developed and tested. To take advantage of innovative technology requires an open mind and a dedication to continued investigation and analysis. It is important not only to stay informed of new control products and techniques as they come into the market, but to review the older procedures implemented throughout the company. In short, to continuously improve the process.

Through continued review and upgrading of ESD control equipment and procedures, costs can be maintained at reasonable levels. The corporation will remain free to concentrate on improving productivity, reducing yield losses, and maintaining the reliability of their products. A commitment to the ongoing improvement of ESD control facilities will also make it easier to protect newer, more sensitive devices when they become a part of the operation.

Points To Remember

The key steps to the efficient design and successful implementation of a custom-tailored ESD control program depend on nothing less than a companywide commitment. The individual steps for setting up or strengthening an ESD control program are listed below.

Step 1 Study and understand the technology of ESD control.

Step 2 Gather scientific evidence of the economic value of ESD control.

Step 3 Establish an active ESD committee.

Step 4 Develop a general plan of implementation.

Step 5 Write or adopt a set of realistic handling requirements.

Step 6 Prepare a detailed statement of policy, including individual responsibilities.

Step 7 Prepare a presentation for management.

Step 8 Establish a top management commitment.

Step 9 Develop the details of ESD program management.

Step 10 Present the program to middle and lower management.

Step 11 Begin implementation of the control facilities.

Step 12 Begin training of the entire work force, including management.

Step 13 Begin using facilities immediately after training.

Step 14 Begin formal auditing of all manufacturing departments.

Step 15 Report auditing results, graphically and with regularity, to all levels of management.

Step 16 Continuously improve the process.

Chapter 3

Fundamentals of Electrostatics

To successfully implement or upgrade an ESD program, familiarity with the concepts of electrostatics and ESD is necessary. In this chapter, we will briefly touch upon the basic elements. For a more extensive discussion, refer to introductory physics or engineering text books.[23-25]

Basic Elements

Electric Charges

Whenever two objects, such as a glass rod and silk cloth, are brought into contact with each other and then separated, the two objects become charged. This charge is readily evident in that the charged objects will then attract bits of paper or cause the hair on your arm to stand up. Furthermore, if two corks are charged by a glass rod, the corks will repel each other. If on the other hand, one cork is charged by the glass rod while the other is charged by the silk, the corks will attract each other. From this simple demonstration, one can conclude that there are two kinds of charges and that like charges repel while opposite charges attract. Another example of this is illustrated in Figure 3-1. Two strips of adhesive tape pulled from a roll have like (positive) charges and repel. The surface of the roll is charged (negative), so the strips are attracted to

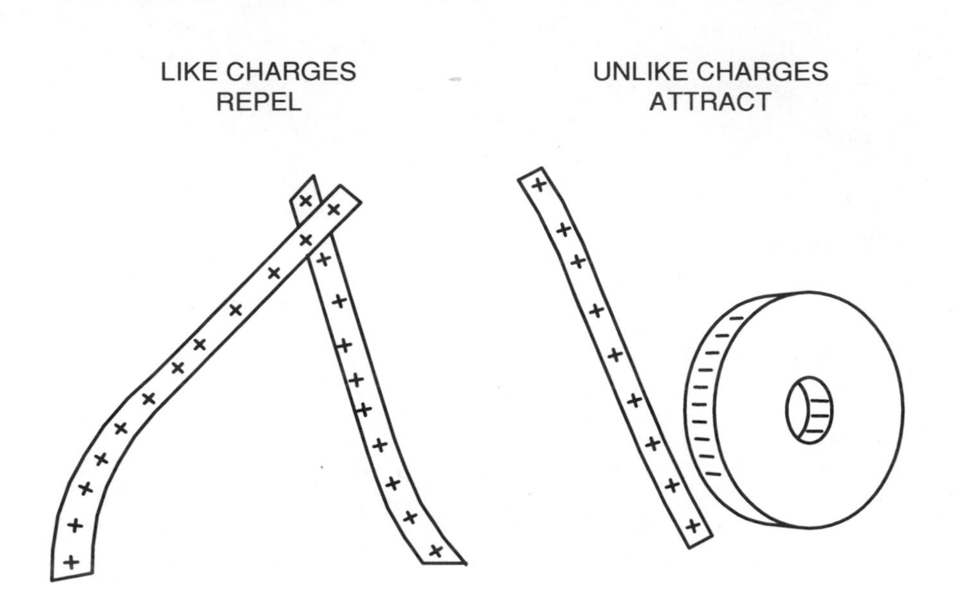

Figure 3-1. Charges generated by unrolling adhesive tape

the roll. The charge on these objects is due to an excess or deficiency of electrons on their surfaces. These charges are referred to as static or static electricity. This is because they can remain stationary on an object for long periods of time. The sudden transfer of this charge from one body to another by oppositely charged bodies being brought into close proximity is called electrostatic discharge (ESD). Lightning is a special, very high energy form of ESD as illustrated in Figure 3-2. The basic unit of charge is the coulomb, and the measurement of these charges on objects is discussed in Chapter 6.

The interactions between electric charges are described by Coulomb's law. Coulomb carefully repeated the glass rod/silk cloth experiment described earlier, measured the force of attraction or repulsion in each case, and found that the force between two small charged objects is:

1. Directed along a line joining the two objects
2. Proportional to the product of the magnitude of the charges
3. Inversely proportional to the square of the distance between them.

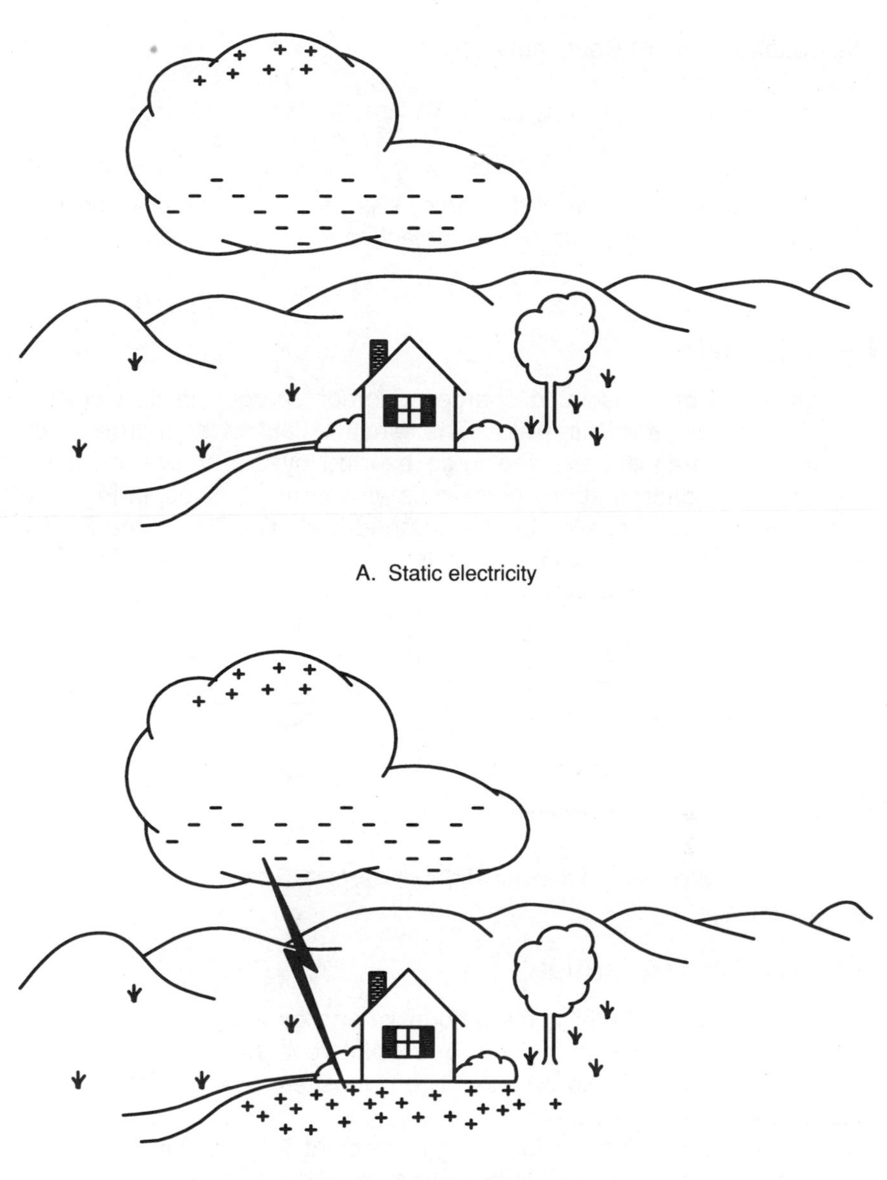

A. Static electricity

B. Electrostatic discharge

Figure 3-2. Lightning is a form of electrostatic discharge.

The equation form of Coulomb's Law is:

$$F = k\,\frac{q_1 q_2}{r}$$

where **F** is the magnitude of the force, **k** is the proportionality constant, q_1 and q_2 are the charges on the object, and **r** is the distant between them.

Electric Field

The influence of an isolated charge on other charges in its vicinity is described by its ***electric field***. The electric field of a charge is the direction and magnitude of the force exerted by the charge on a unit (one coulomb) charge at any point in its environment, Thus, in Figure 3-3, the charge **+q** is repelled by the charged body **Q** with a force **F**. The field intensity of the object **Q** is just **F/q**.

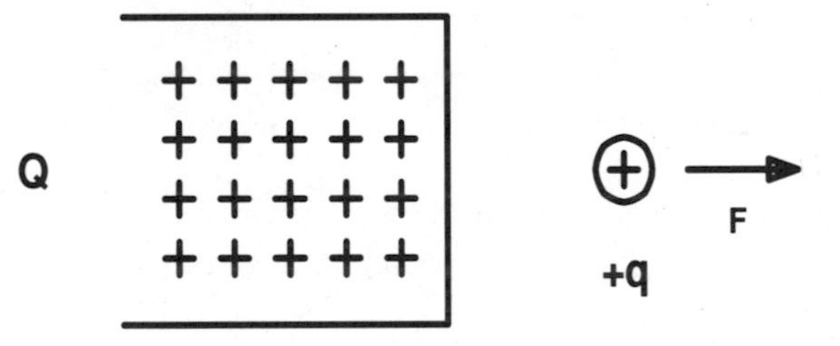

Figure 3-3. The force on a point charge from a charged body

Electrostatic Potential

The electrostatic potential is the amount of energy (work) per unit charge required to move a charge from one point to another in a field. For example, the electric field between two oppositely charged metal plates is represented in Figure 3-4.

The movement of the charge **+q** from point P_1 to point P_2 requires work since the electric field (force) wants to move the charge toward the negative plate. The work done in moving this charge is:

$$W = FX = qEX$$

or

$$W/q = EX$$

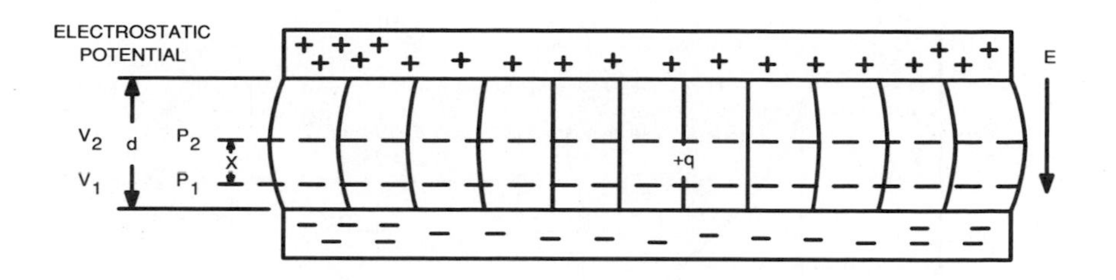

Figure 3-4. The work done in moving the charge +q from P_1 to P_2 is $Q(V_2 - V_1)$

where **E** is the field strength and **X** is the distance moved. It is customary to speak of this work per unit charge as a potential difference, $V_2 - V_1$. Knowing this relationship allows one to create a known electric field between the plates by using an external voltage source. In that case,

$$E = V(plates)/d$$

Capacitance

The parallel plate configuration in Figure 3-4 is an example of a ***capacitor***. The ratio between the charge on the plates and the potential difference between the plates is the ***capacitance***, **C**.

$$C = Q/V$$

This definition is general, regardless of the configuration, size, or shape of the elements of the capacitors. The value of **C**, however, is highly dependent on the plate size and shape. In the case of the parallel plate capacitor, the capacitance is given by:

$$C = \varepsilon A/d$$

where **A** is the area of the plates, **d** is the spacing between them, and ε is the ***dielectric constant*** of the material between them. The dielectric constant of air is very close to unity and for most insulators varies between 1 and 10. This relationship can be used to great advantage in designing protective packaging as discussed in Chapter 12. The human body has a capacitance of about 50 to 500 pF. This capacitance depends on many variables, including shoe sole thickness (Figure 3-5). A standard value of 100 pF has been adopted for testing semiconductor devices for their sensitivity to ESD (see ESD Failure Models). A circuit board tote box may have a capacitance of about 50 pF, and a packaged integrated circuit one of only a few pF.

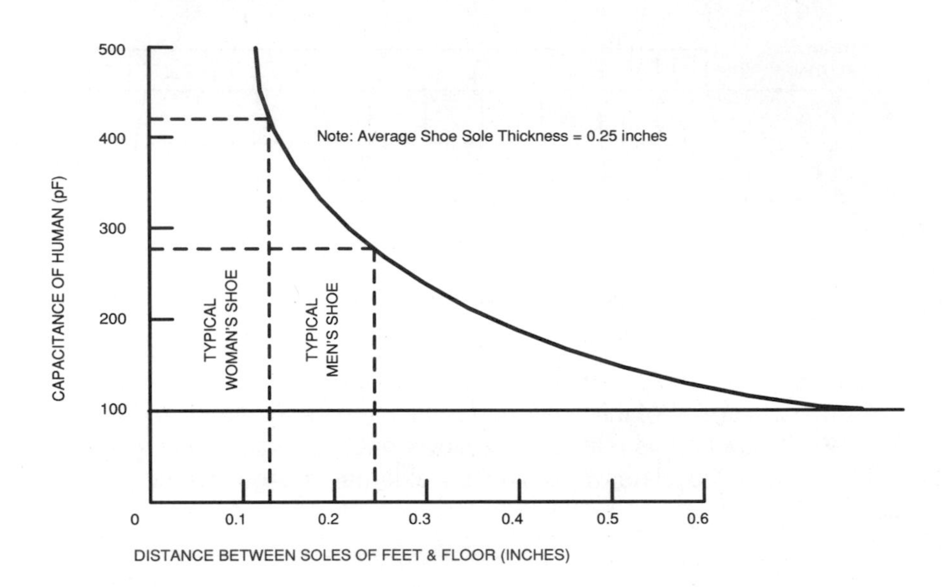

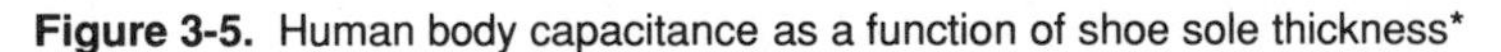

Figure 3-5. Human body capacitance as a function of shoe sole thickness*

** Printed with permission from Interference Control Technologies, Inc./Don White Seminars. Copyright protected.*

Another important quantity is the energy stored in a capacitor. This is given by:

$$E = 1/2\ CV^2 = 1/2\ Q^2/C$$

Thus, a typical human body with C = 100 pF charged to 5000 volts will have available $1/2(10^{-10})\ (5\times10^3)^2$ = 1.25 millijoules of stored energy. This is more than enough to damage a small semiconductor device.

Charge Generation

In the subsection on electric charges, we mentioned the generation of a charge involving glass, silk, and cork. If charge generation were limited to these few materials and situations, ESD would not be a major problem.

In general, whenever two materials, at least one of which is an insulator, are brought together and then separated, there will be a flow of electrons from one material to the other. The material giving up electrons becomes positively charged while the material accepting the electrons becomes negatively charged. The relative tendency to become positively or negatively charged is an intrinsic material property.

This property is summarized in the triboelectric series (Table 3-1). The term, ***triboelectricity***, refers to charging that occurs as a result of contact and frictional motion. However, in common usage, the terms contact charging and tribocharging are used more or less interchangeably. The triboelectric series, in general, does predict the correct polarity (sign) of charge observed in either case. We will not discuss the theory of contact charging; an excellent review of the literature available is given in the References.[26]

While contact charging is the most common way that a static charge is developed, other processes, such as ion beam charging, spray charging, photoelectric charging, and corona charging, are also potential sources of static charges. In the manufacturing environment, taping, operations involving conveyor belts, and any automated process can include charge-generating processes.

When the charged object is an insulator, the charge may persist for long periods of time because insulators (by definition) are poor conductors of electricity. This is where the term static charge comes from. Because of this difficulty in removing a charge from insulators, it is preferable that the charge not be generated at all. Treating insulators to make them antistatic is discussed in Chapter 12.

The amount of charge that appears on a surface is the net result of competing charge generation and discharge (dissipation) processes. Some factors which affect these processes are summarized in Table 3-2. The relative position of the material in the triboelectric series is only one factor in the charge generation process. Two materials that are close in the table may generate a large static charge if the pressure of contact is high. On the other hand, materials well separated on the chart may generate a relatively small static charge if their contact mechanics are affected by ambient relative humidity (Figure 3-6). A static charge can be characterized by measuring the charge itself (using a coulombmeter) or by measuring the electrostatic voltage or field of the charge. These measurement techniques will be reviewed in Chapter 6. In general, it is easiest to compare the charge generation mechanisms by the amount of voltages they generate. The electrostatic voltages generated by several different events are summarized in Table 3-3.

While the ambient humidity does have a significant effect on these voltage levels, damaging voltages can still be generated, even at 55% relative humidity or higher.

Table 3-1. Triboelectric Series

MATERIALS	POLARITY (+ OR –)
Asbestos	Acquires a more positive charge
Acetate	
Glass	
Human hair	
Nylon	
Wool	
Fur	
Lead	
Silk	
Aluminum	
Paper	
Polyurethane	
Cotton	
Wood	
Steel	
Sealing Wax	
Hard Rubber	
Acetate Fiber	
MYLAR*	
Epoxy Glass	
Nickel, Copper, Silver	
UV Resist	
Brass, Stainless Steel	
Synthetic Rubber	
Acrylic	
Polystyrene Foam	
Polyurethane Foam	
SARAN†	
Polyester	
Polyethylene	
Polypropylene	
PVC (vinyl)	
TEFLON*	
Silicone Rubber	Acquires a more negative charge

* Trademark of E. I. Du Pont de Nemours & Co., Inc.
† Trademark of Dow Chemical U.S.A.

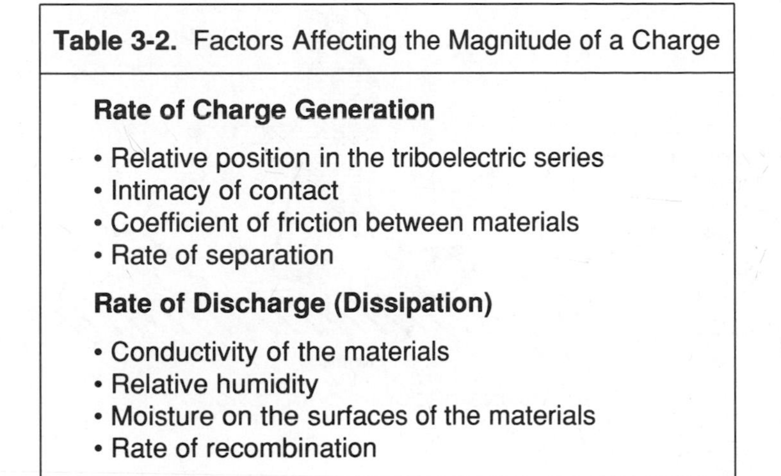

Table 3-2. Factors Affecting the Magnitude of a Charge

Rate of Charge Generation

- Relative position in the triboelectric series
- Intimacy of contact
- Coefficient of friction between materials
- Rate of separation

Rate of Discharge (Dissipation)

- Conductivity of the materials
- Relative humidity
- Moisture on the surfaces of the materials
- Rate of recombination

Table 3-3. Typical Electrostatic Voltages (Volts)

EVENT	RELATIVE HUMIDITY		
	10%	40%	55%
Walking across a carpet	35,000	15,000	7,500
Walking across a vinyl floor	12,000	5,000	3,000
Motions of bench employee	6,000	800	400
Removing dual in-line packages (DIPs) from plastic tubes	2,000	700	400
Removing DIPs from vinyl trays	11,500	4,000	2,000
Removing DIPs from polystyrene foam	14,500	5,000	3,500
Removing bubble pack from PWBs	26,000	20,000	7,000
Packing PWBs in foam-lined box	21,000	11,000	5,500

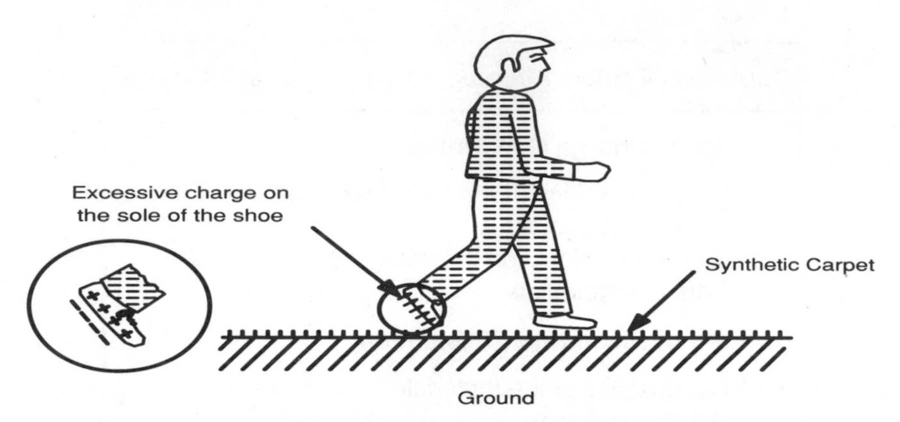

CHARGING PROCESS OF A PERSON

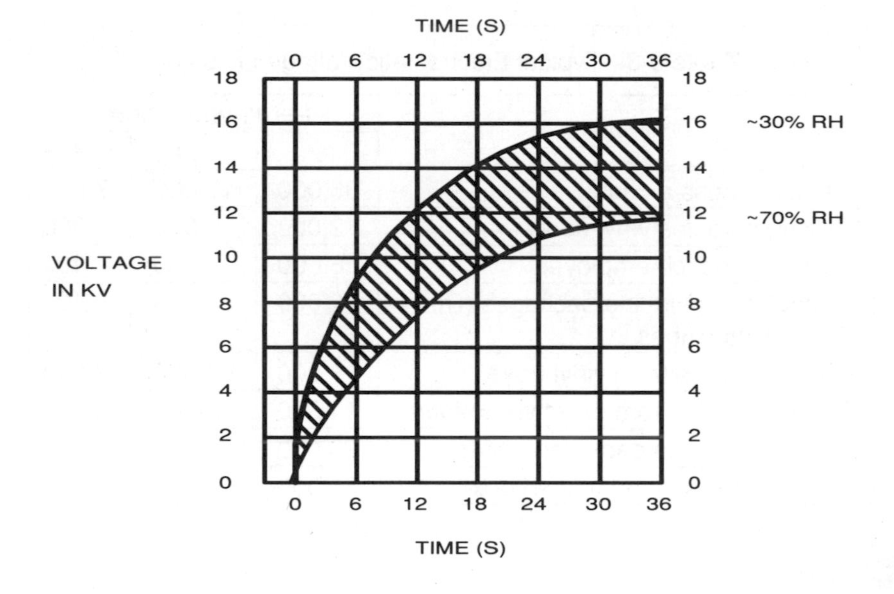

GRAPH OF VOLTAGE BUILD-UP

Figure 3-6. Charge build-up on a person at different relative humidities*

** Printed with permission from Interference Control Technologies, Inc./Don White Seminars. Copyright protected.*

ESD Failure Models

A wide variety of models have been proposed to represent ESD events. In general, these models fall into two categories. The first is the human body model (HBM). A simple equivalent circuit model is given in Figure 3-7A. This model should also include a series inductance, but most test methods regard such inductance as parasitic. Suggested values for the resistance (100 to 10K ohms) and capacitance (50 to 500 pF) vary. A typical HBM pulse has a rise time of less than 10 nanoseconds and a decay time constant of 50 to 300 nanoseconds. An idealized HBM current waveform is given in Figure 3-8A. The second major model is the charged-device model (CDM). The CDM equivalent circuit model is given in Figure 3-7B. The model represents the ESD event that occurs when a device becomes charged due to some manufacturing process, such as sliding down an integrated circuit (IC) shipping tube or random motion in a tape-and-reel carrier tape. When the device is subsequently grounded, a rapid discharge occurs. This discharge current is limited only by the parasitic impedance and capacitance of the device. As a result, rise and fall times for the CDM are typically less than one nanosecond. An idealized CDM current waveform is shown in Figure 3-8B. The HBM and CDM events tend to produce different types of failure and require different types of control and protection. Figure 3-9 contains photomicrographs of HBM and CDM failures in the same type of device.

Early in the development of manufacturing processes for microelectronic circuitry, HBM events were the primary cause of ESD yield losses in factories. In recent years, personnel grounding has become a routine part of manufacturing, and most devices have at least some protection from HBM events. As a result, HBM failures in manufacturing have decreased significantly.[27] However, the HBM is still a major cause of failures in the field. Handling PWB assemblies containing sensitive devices during repair and maintenance tends to be less controlled. Furthermore, HBM ESD and similar events are major causes of operating equipment malfunctions.

On the other hand, the CDM is just emerging as a significant ESD concern.[28-32] This is due to the rapid changes occurring in circuit assembly procedures. Many new processes are being introduced into factories without sufficient regard for static generation. Improperly designed or installed conveyor belts, slides, flow racks, robotic arms, or test equipment can systematically damage CDM-sensitive devices.

One particularly subtle and common way in which CDM occurs is by static induction.[28,33] This series of events is illustrated in Figure 3-10. In this process, a neutral object (such as an integrated circuit) is placed near a static charge residing on an insulator in the work area.

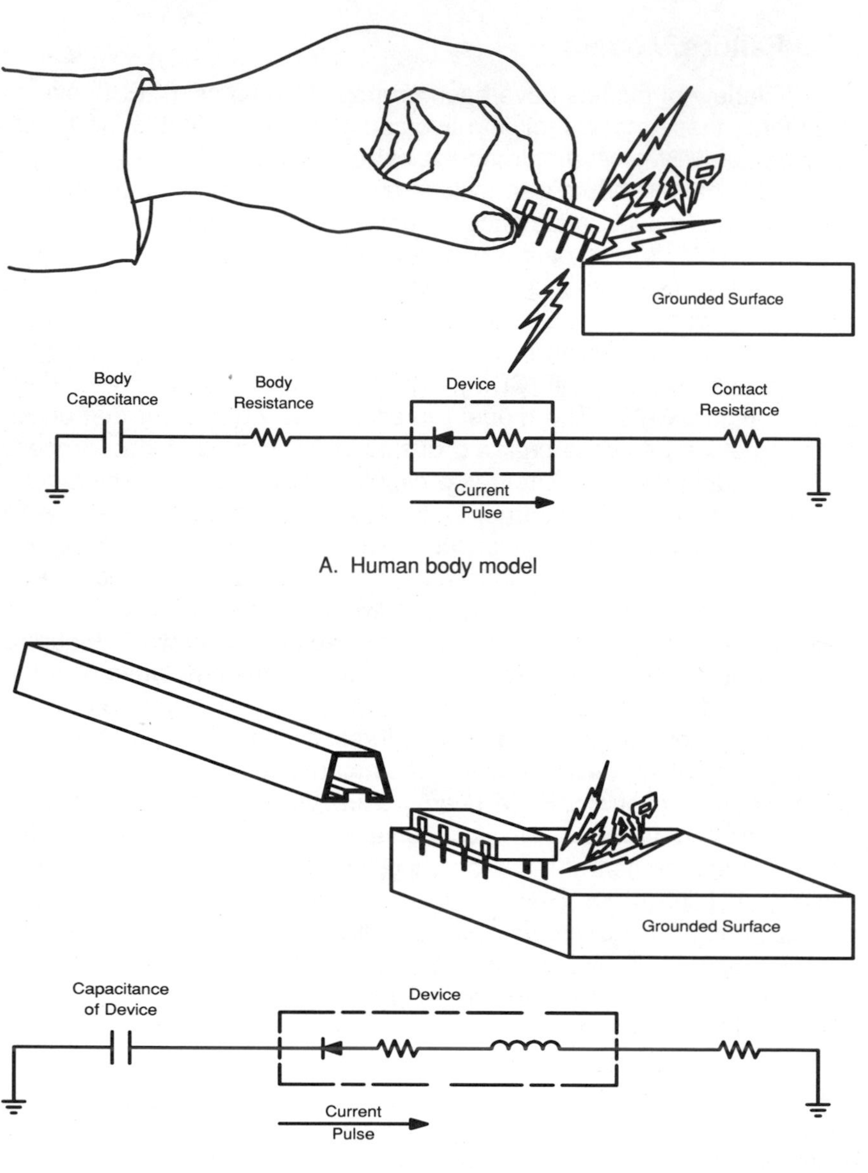

Figure 3-7. Simple equivalent circuit models

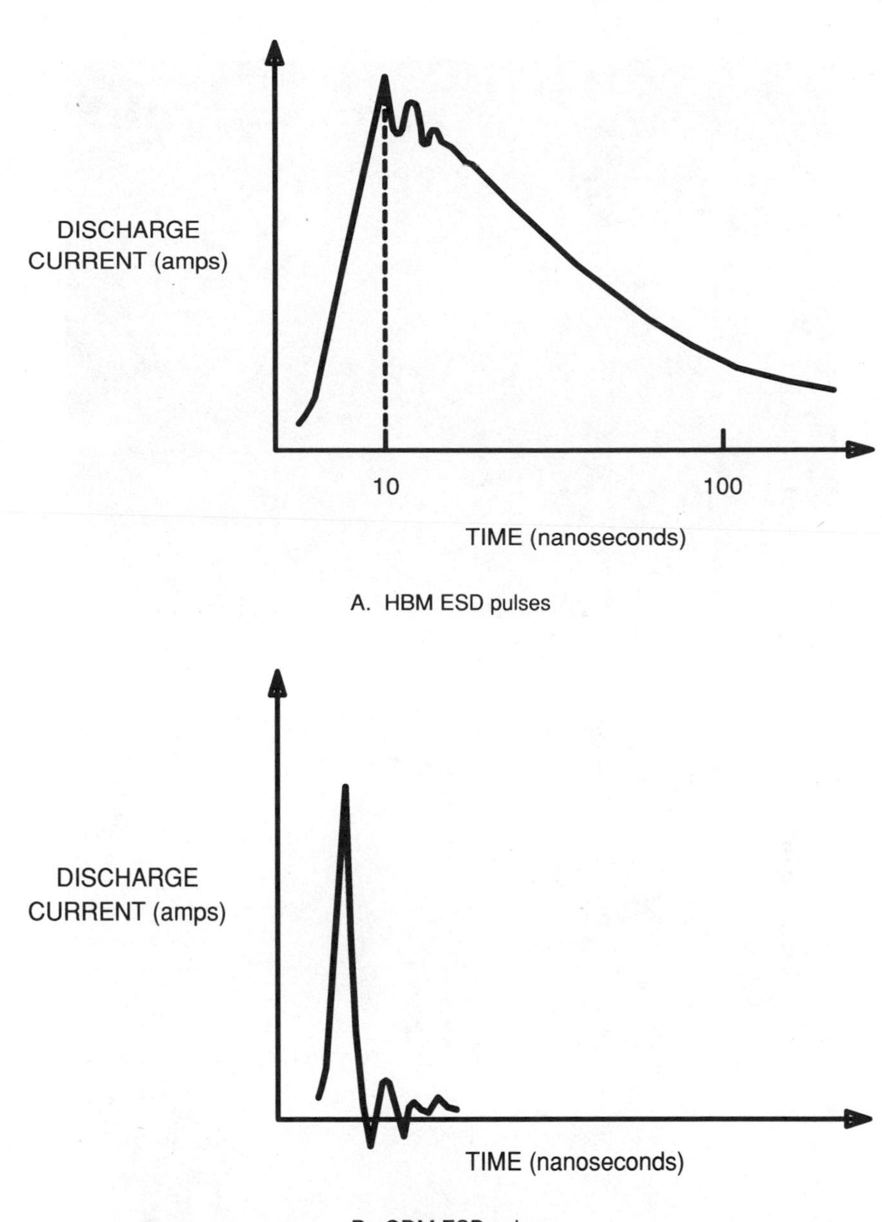

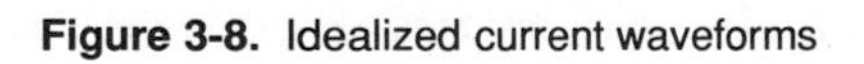

Figure 3-8. Idealized current waveforms

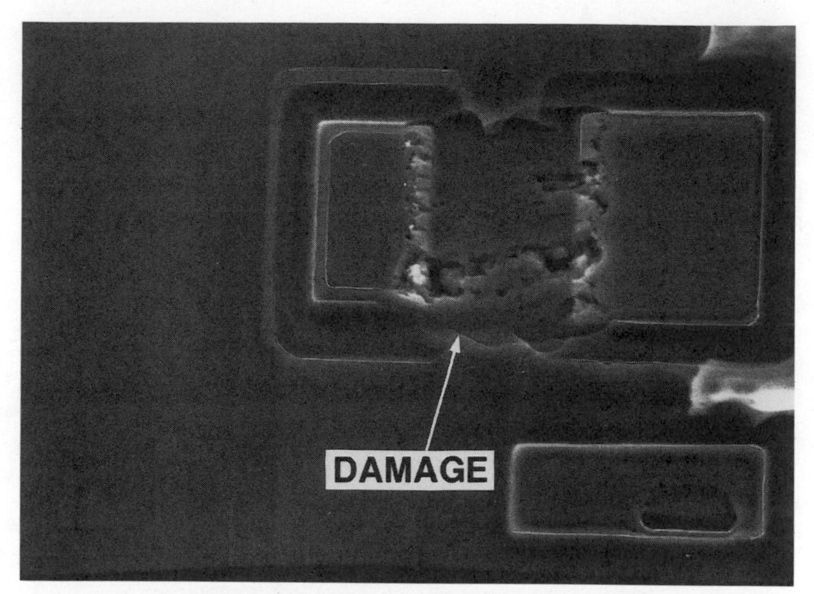

A. HBM causes melting of silicon in pn junction area

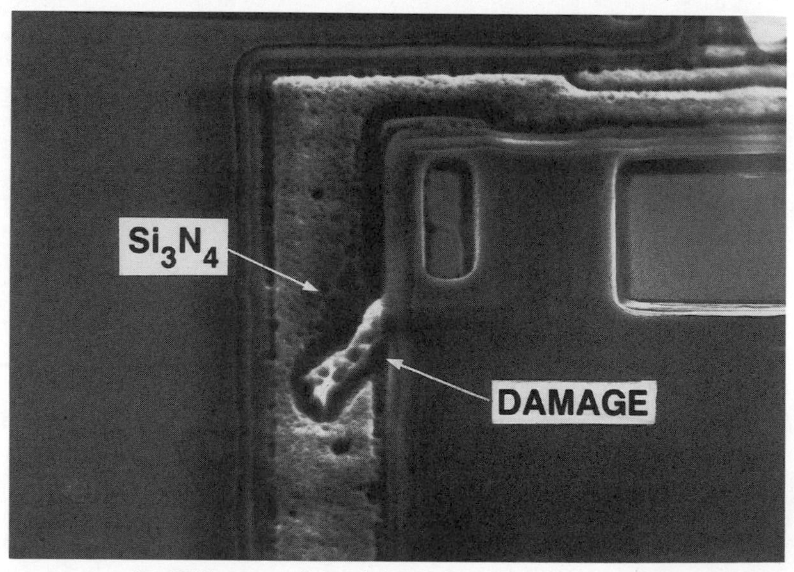

B. CDM causes dielectric breakdown of Si_3N_4 insulation

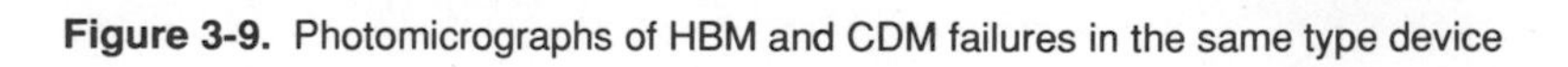

Figure 3-9. Photomicrographs of HBM and CDM failures in the same type device

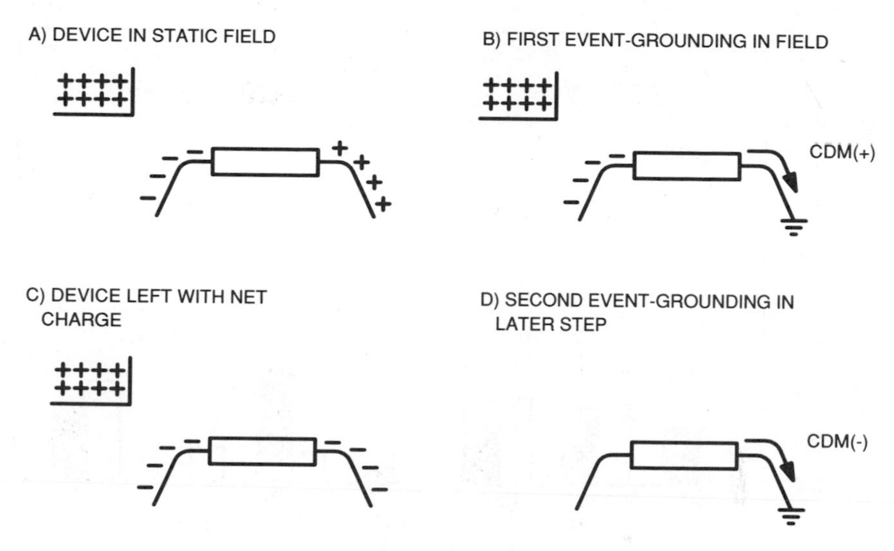

Figure 3-10. ESD by induction

The field resulting from this static charge induces a potential on the device. This is indicated in Figure 3-10A as a charge separation on the lead frame of the device. If one of the device leads is then grounded, a current will flow to bring the device back to zero electrostatic potential (ground). The threat to the device does not end there. The object now has a net residual charge until that dissipates into the air, which may take minutes or hours, or the device is grounded in a subsequent process step. If the latter occurs, the result is a second event, opposite in polarity and equal in magnitude (charge) to the previous one.

These models are used as the basis for standard test methods for the characterization of devices and systems. It has become customary to express the relative susceptibility of items to ESD in terms of the voltage that appears on the capacitor in these models. The ***withstand voltage*** of an item is defined as the highest voltage that the item is known to be able to withstand without changing its operating characteristics. The range of sensitivity of microelectronic devices is from about 20 volts for unprotected metal oxide semiconductor (MOS) transistors to greater than 15,000 volts for some Zener diodes. MOS was an early focus of ESD protection because of the inherent sensitivity of the thin (<1000 Angstroms) gate oxide. It became popular to state that MOS integrated circuits were more sensitive than bipolar devices. However, data recently collected by the military suggest that device technology is a poor indicator of device sensitivity (Figure 3-11).

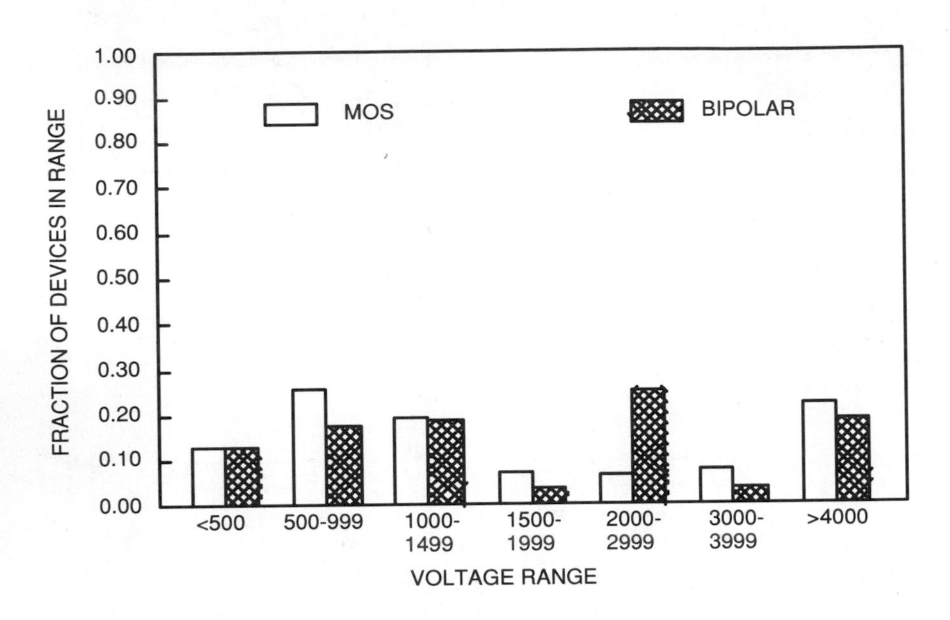

Figure 3-11. HBM ESD withstand thresholds of MOS and bipolar devices

Therefore, control procedures are the same for both types and will depend more on the actual sensitivities of the device being handled, rather than the type of device. It is therefore important that all new device designs be tested for their HBM and CDM sensitivity so that proper handling precautions can be applied.

Charge Removal

From the preceding paragraphs, it is clear that static charges are the primary cause of device failure (first step) in the various ESD scenarios. They are represented by the charged capacitors in the HBM and CDM failure models. Thus, safe removal of these charges can greatly reduce device failures. Removing these charges from personnel or preventing charge accumulation is accomplished by the familiar grounded wrist strap and can be done because the human body is a relatively good conductor.

Charges on insulators would not be a problem if it were not for the phenomenon of induction charging described in the previous section. The only reliable method of removing a charge from insulators is ***air ionization***. Air ionizers use high-voltage or nuclear sources to produce positive and negative ions, which are then introduced into the vicinity of potential static generators. These ions will combine with the charges on the surface of the insulator, effectively returning the insulator to zero potential.

The application of air ionizers is most effective when a specific charge generation process has been identified in a manufacturing sequence. Air ionization can then be employed to neutralize these charges as they appear.

Charge Dissipation and Protective Materials

In the CDM model, the threat to the device can be greatly minimized by providing a moderate resistance in the discharge path. In practice, this resistance is provided by the materials that compose those objects that the device is likely to contact in the manufacturing environment, such as bench tops. These materials will be discussed in Chapter 12. They are most commonly evaluated according to various applicable resistance parameters such as point-to-point resistance, resistance-to-ground, volume resistivity, and surface resistivity. As the resistivity parameters are often misunderstood, we will review their definitions here.

The volume or bulk resistivity, ρ_v, is defined for a homogeneous material as:

$$\rho_V = Rwt/l$$

where **R** is the measured resistance of a sample of length l (that is, the spacing between the measurement probes), of width **w**, and of thickness **t**. The units of ρ_v are in ohm-centimeters.

Surface resistivity, ρ_s, is used to characterize materials that primarily allow charge dissipation along their surfaces.

If measurement electrodes are placed on the surface of a **w** by l piece of material as in Figure 3-12, the resistance and resistivity are related by:

$$R_s = \rho_s l / w$$

If **w** equals l, that is, the specimen is square, then:

$$\rho_s = R_s$$

Thus, ρ_s is the resistance of a square piece of material (provided that the electrodes extend fully along the opposite edges of the material). Furthermore, all squares, no matter what their size, have the same resistance. For this reason, ρ_s is reported as ohms/square or $\Omega/\square$.

Consider a sample such that l = **5w**. This sample can be viewed as being composed of 5 **w x w** squares (Figure 3-13).

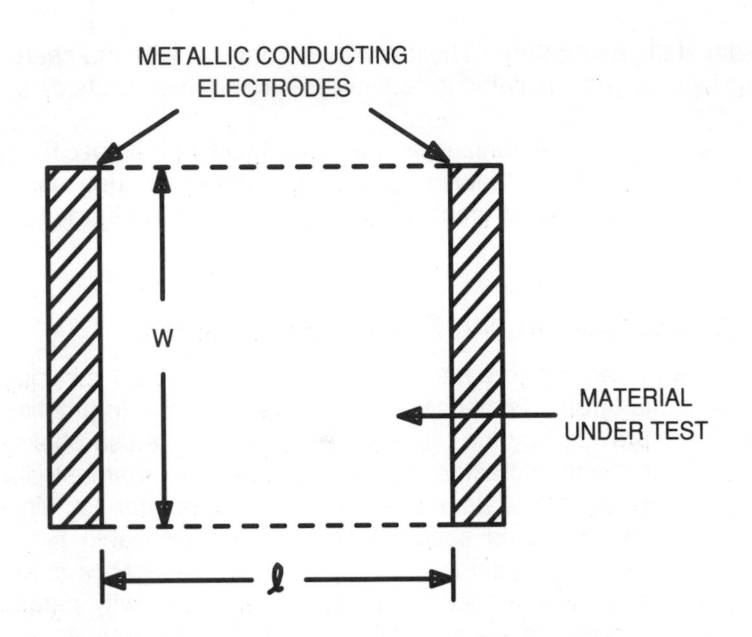

Figure 3-12. Electrode configuration for measuring surface resistance

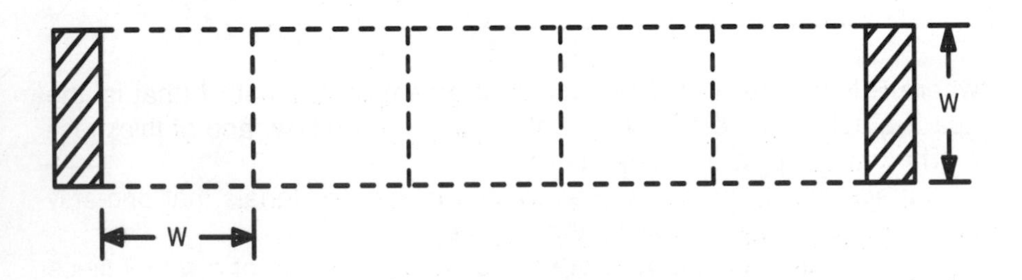

Figure 3-13. Counting squares in a rectangular sample

The resistance of this sample is:

$$R_s = \rho_s 5w/w = 5\rho_s$$

The resistance is proportional to the number of squares with edge length **w**.

Points To Remember

- Static charges are generated by materials coming into contact with each other and then separating.
- Charges on insulators can remain for extended periods of time and can only be fully removed by air ionization.
- The electric field of a static charge can induce potentials on conductors and devices.
- It is easy to generate a static charge that can damage many electronic devices.
- Static charge accumulation can occur at any stage of manufacturing, installation, repair, and operation.
- There are two main ESD failure models, the human body model and the charged-device model.
- The ability of an object to store a charge is given by its capacitance, which is a strong function of shape, size, and material composition.

Chapter 4

An Economic Analysis

It has now been established that ESD can damage virtually any semiconductor. However, establishing how often that occurs in a specific manufacturing facility, or the consequence of that damage, has proven to be extremely difficult. For that reason, approvals to expend funds for ESD precautions are sought reluctantly and are often denied. This chapter will illustrate a technique that has proven successful in estimating the economic benefit of ESD precautions and, subsequently, in establishing a systematic and cost-effective prevention plan.

Essentially, this approach consists of conducting carefully controlled experiments in a manufacturing environment so as to provide comparisons with and without ESD precautions on given production lines. The results of these experiments will make it possible to justify the general use of ESD precautions on the same production lines, thereby establishing the opportunity to evaluate the impact of ESD precautions on manufacturing. Manufacturing data is then gathered and compared with the experimental data. This combination of experimental and manufacturing data should provide a strong argument for the application of ESD precautions throughout all manufacturing facilities.

For companies starting their own ESD program, the five case studies presented here will provide a solid base of irrefutable economic evidence on which to build a business case. These studies also provide convenient examples for the development of additional experiments on other production lines. ***This experimental technique provides a scientific means of not only assessing the impact of ESD damage but also of evaluating which control techniques will be effective and which will be overkill.***

Additional experimental data and case studies may be found in much of the published ESD literature, especially in the proceedings of the EOS/ESD Symposia. The additional data and case studies combined with the experimental data presented here, or the results of any proprietary experiments, make a strong case with which to enlist broad based management support for a corporate commitment to ESD control. As ESD control programs are implemented in different manufacturing facilities, the information provided by these experiments can be used to form the basis for realistic handling requirements and for auditing programs that verify across-the-board compliance.

The five case studies are divided, by product line, into three main topics. Case Study 1 (Resistor Failure) briefly documents an isolated experiment involving PWB assembly. Case Studies 2, 3 and 4 focus on the manufacture of components and PWB assemblies and combine experimental data with actual manufacturing data. Case Study 5 demonstrates the difficulty in developing satisfactory control techniques for handling ultrasensitive devices.

Case Study 1: Resistor Failure Due to Automation in Production

Internal customer complaints from an AT&T manufacturing location in North Carolina prompted a study of damaged resistors. The resistors involved were thin film integrated circuit precision resistors, specified to be within a tolerance of ±0.2 percent. However, rather than being precise well-controlled resistors, they demonstrated values actually four to five times higher than the intended value. Customer complaints were certainly justified and worthy of close inspection.

The damaged resistors are illustrated in Figure 4-1. The resistor on the right shows lightning-like damage appearing as lines running across the material in the pattern of a "crow's foot." The dark area in the photograph is the tantalum nitride that makes up the resistor material. The solid light lines in the tantalum nitride are created by a laser. The laser is used to trim the resistor to a predetermined value with a tolerance of 0.2 percent. The resistor on the left demonstrates clean laser-generated lines and measures 300 ohms. The resistor on the right, with obvious damage sites, measures 1411 ohms.

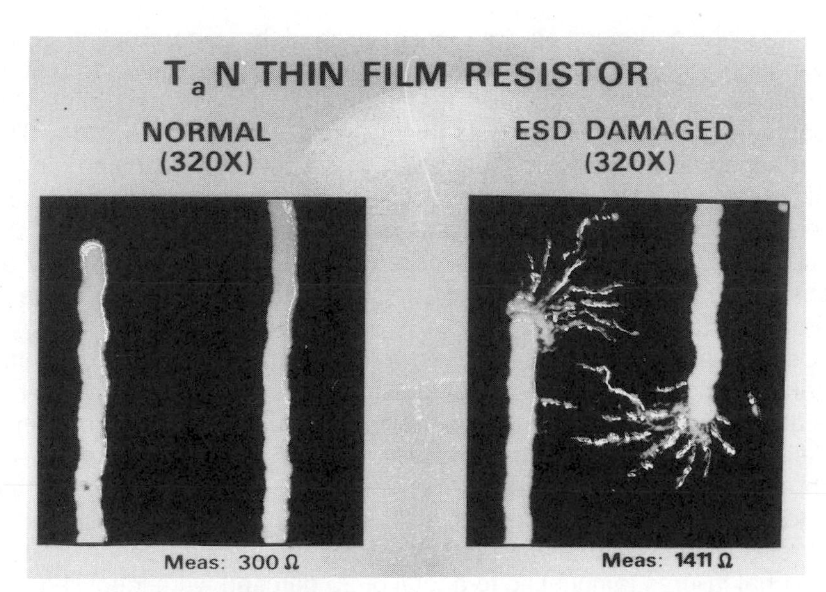

Figure 4-1. ESD damaged thin film resistor — "Crows Foot"

In searching for the cause of this damage, ESD was identified as one possible cause. To prove the hypothesis, the damage was duplicated with an ESD transient of 450 volts, using the charged-device model. This relatively low threshold of damage to resistors raises many questions, including that of the effect of ESD on silicon devices. A study was initiated to calculate the magnitude of the resistor problem on the manufacturing floor.

Experiment 1

On-site, the average dropout rate for resistors was 20 percent, but some shipments had up to 80 percent losses. In order to determine where the resistors were being damaged, they were tested at each stage of the production line. No failures were detected until the mass soldering machine. In fact all devices were proven to be in perfect working order prior to their immersion in the mass soldering machine. Boards were then tested at several different points throughout the mass soldering operation.

The mass soldering machine consists of a fluxing station, a soldering station, and a cleaning station. Resistors were tested before and after each of these operations. The damage was determined to be occurring during the cleaning operation.

The cleaning operation was accomplished by nine rotating nylon brushes in a solvent bath. The solvent was highly resistive (1,000,000 megohm-cm) and was actually instrumental in allowing the battery of brushes to charge the boards as they passed through the station. The boards then discharged into the ground of the soldering machine. In the dark, even a casual observer could actually see arcing taking place. There was, in reality, a small lightning storm taking place inside the soldering machine as the boards passed through. As measured by an electrostatic locator, there were charges of up to 6000 volts on some of the boards. (See Figure 4-2.)

Using the standard test for defect levels at the conclusion of the manufacturing process, the tester determined that the dropout level of the boards was only 10 percent. That was a surprisingly low incidence of failure in light of the fact that every single PWB assembly was abused with at least nine discharges. It may be taken as an indication of the random behavior of static electricity. However, through tests designed to find resistance shifts that were not caught by the standard test methods, additional failures (amounting to a total of 20 percent) were identified.

The remedy to the problem was found in the choice of the cleaning solvent used in the machine. A more conductive solution was necessary (less than 10,000 megohm-cm) to allow the charge to bleed off as quickly as it was created, without the violent discharges taking place. The soldering facility has since been converted to use an aqueous cleaning solution that virtually eliminates the charging.

	North Carolina	**Merrimack Valley**
No. of Circuits	100	100
Failures detected after	Mass soldering	Mass soldering
Percent of Defects	20	1
Static Measurements	6000V	1500V
Visible Arching	yes	no
Cleaning solvent	10^{12} Ω - cm	10^{11} Ω - cm
No. of Brushes	9	3

Figure 4-2. Two experiments indict PWB cleaning as the source of ESD damage to resistors in 1978.

Experiment 2

Based on the experience with mass soldering machines gained in Experiment 1, a study of the manufacturing facility in North Andover was undertaken. This study was instigated by the discovery of resistor damage on another device (Figure 4-3), but this time at the AT&T Merrimack Valley facility in North Andover, Massachusetts. Again, measurements were taken at each stage of the manufacturing process, with particular attention given to the mass soldering operation. The damage is less severe than that seen at the North Carolina facility, as the charging level was 1500 volts versus 6000 volts (Figure 4-2).

As in the earlier experiments, no damage was observed prior to the resistors passing through the mass soldering machine. Within that operation, it was observed that damage occurred during the time that resistors passed through the cleaning station. However, unlike the North Carolina plant, where dropout rates as high as 80 percent were found, a failure rate of only 1 percent was found. Also, the resistance shift out of tolerance was less, only 118 ohms from the specified value of 406 ohms to 624 ohms.

There are two differences between the cleaning machines at North Carolina and Merrimack Valley. First, where the machines in the southern plant had nine brushes, those in the north had three. Second, where the cleaning solvent in North Carolina was highly resistive, the cleaning solvent in the Massachusetts plant was more conductive (100,000 megohm-cm), but not sufficiently conductive to prevent charging.

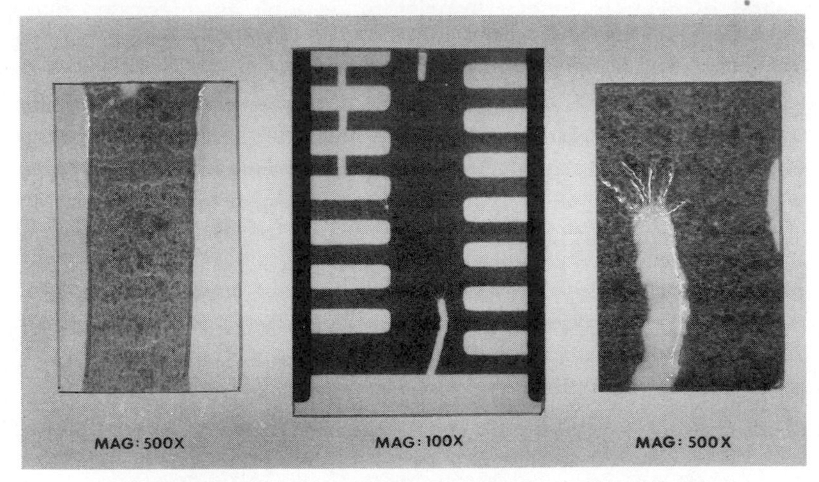

Figure 4-3. ESD damaged thin film resistor — "Crows Foot"

Consequently, the static potential was measured at approximately 1500 volts instead of 6000 volts. Accordingly, a much lower failure rate was noted among the resistors. Furthermore, additional measurements of resistance produced no evidence of additional failures, as had been the case with the earlier testing. The soldering facility has since been converted to use an aqueous cleaning solution that virtually eliminates the charging.

Case Study 2: Bipolar Discrete Device Failure

The experiments discussed in both this and the following case study were accomplished by randomly selecting samples of the product, splitting these samples into two equal populations, implementing the desired ESD precautions on only one population, and, finally, processing them simultaneously through the manufacturing process. Therefore, the only variables in the experiment were the ESD precautions, thus allowing a scientific look at the impact of these techniques.

Experimental Evidence

Case Study 2 presents an isolated experiment, selected to illustrate that ***even older product lines using bipolar devices can benefit*** from using wrist straps during manufacturing. The experiment was conducted in 1979 on a PWB assembly line, using as a test vehicle a PWB assembly that had been initially introduced in 1973. This PWB assembly consisted of discrete bipolar transistors as well as integrated circuits. The only form of ESD protection was the introduction of wrist straps to one of the two populations. The sample size for each cell was 216 PWB assemblies; the relative humidity averaged 20 percent.

The test results showed significant differences in the two populations, with the unprotected population having 2.7 times the number of test defects than the protected population (see Figure 4-4). The defective components were misplaced before a failure mode analysis could be conducted.

In conclusion, the significant improvement (2.3 percent failure versus 6.2 percent) in test defect rates strongly suggests that using wrist straps was beneficial in the manufacture of this older bipolar product. However, without the diagnostics, a final conclusion was difficult to arrive at. Therefore, another experiment was conducted using a different product.

	ESD Protected (Note)	Unprotected
Final Test Defects	2.3%	6.2%
Relative Humidity	20.0%	20.0%

(Lot Size = 216 PWB assemblies)

Note: Protection consisted of wrist straps only.

Figure 4-4. Results of PWB assembly experiment conducted in 1979 on a product initially introduced in 1973

Case Study 3: Device and PWB Assembly Failures

This case study represents a particularly complete collection of both experimental and manufacturing data. It is a study of the ESD behavior of a bipolar Hybrid Integrated Circuit (HIC) during manufacture, as well as during PWB assembly and test.

The following studies were initiated as a result of external customer complaints from telephone companies and unusually high failure rates. It was determined that ***PWB assemblies were failing on customer premises due to ESD damage***. A bipolar junction on HIC "A" on the PWB assembly was exhibiting excessive leakage, resulting in system disruption. The junction damage encountered is shown in Figure 4-5 in a photograph of a field failure.

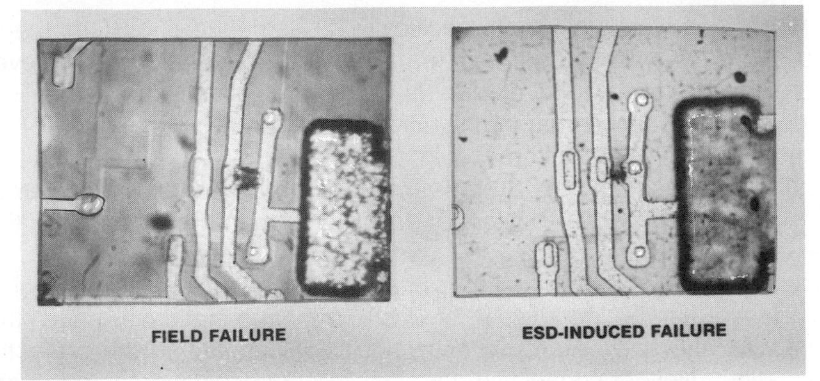

Figure 4-5. Comparison of a field failure (1000x) and an ESD-induced failure (1000x) of a bipolar junction

The ESD-induced failure reveals nearly identical damage to that resulting from ESD testing using the HBM.[34,35] The threshold of damage was established to be 700 volts.

Experiments were conducted in the HIC manufacturing shop and in a PWB assembly shop. After ESD precautions were introduced, additional data was drawn from the two shops. While manufacturing data by itself may be highly suspect due to the large number of variables involved in the manufacturing process, definitive conclusions can certainly be drawn when manufacturing evidence is supported by experimental data.

Experiment 1

This experiment was conducted in the HIC manufacturing shop where the relative humidity is regulated between 30 and 50 percent year-round. The selected test vehicle was HIC "A," described above. The ESD precautions, unique to the protected population, consisted of wrist straps and static dissipative tote trays. The sample size was 1275 HICs in each group.

As part of the manufacturing process, both populations were tested twice: once at an in-process test station and again at final test. The results were nearly the same at both test operations, with the unprotected HICs yielding significantly more defects than the protected population (Figure 4-6). During the in-process testing, the unprotected population had a defect rate of 4.5 percent, compared to 2.4 percent for the protected population, or 1.9 times the number of electrical defects. In the final test, the unprotected population had a defect rate of 2.9 percent versus 1.2 percent for the protected HICs, or 2.4 times as many defects.

All defects were analyzed specifically for evidence of ESD damage, and none were found in the protected population. However, approximately half of the defects in the unprotected group were ESD induced, totally accounting for the difference in the two populations.

In conclusion, the difference in the defect rates between the two populations was found to be ***statistically significant with a confidence level of 99.9 percent***. Additionally, failure mode analysis confirmed that the difference was due to ESD damage. Therefore, by virtue of the design of the experiment, it can be assumed with the same level of confidence that the introduction of wrist straps and static dissipative tote trays was effective in the prevention of ESD damage to the protected population. Neither conductive nor shielding materials were used at any point during this experiment.

Experiment 2

Because the results had been so definitive in the HIC shop, it was decided to extend the experiment into the PWB assembly shop and to design the experiment to determine whether the ESD precautions taken in the HIC shop had influenced the quality of the outgoing HICs. This was accomplished by marking all of the good HICs in Experiment 1 such that the parent population could be determined later. The HICs were then transported between the two manufacturing shops with the protected HICs in the new conductive foam and the unprotected HICs in the old expanded polystyrene (EPS) shipping trays. The use of EPS trays was formerly standard procedure but has since been replaced by conductive foam. The HICs were mixed and randomly introduced into the PWB assembly manufacturing process and allowed to proceed without supervision, with one exception. Employees at test positions were instructed to pull out any defects with the unique markings.

It is important to note that, at this time, the PWB assembly shop was not employing any form of ESD protection.

The results revealed that the defect rate of the HICs that had been unprotected in the HIC shops was 5.5 times greater than the defect rate of those that had been protected. Several HICs were damaged during removal from the PWB assemblies, and therefore, failure mode analysis was not possible. Figure 4-6 summarizes the results for both Experiment 1 and Experiment 2.

In conclusion, the difference in defect ratio of 5.5 to 1 was found to be ***statistically significant with a confidence level of 90 percent***.

	ESD Protected	Unprotected	Ratio
HIC shop:			
First Test Defects	2.4%	4.5%	1.9
Final Test Defects	1.2%	2.9%	2.4
Relative Humidity	35.0%	35.0%	
Continuation—EQ Shop			
(Without Protection)			
Defect Rates	0.4%	2.2%	5.5

(Lot Size = 1275 HICs, each group)

Figure 4-6. Results for HIC Experiments 1 and 2

Therefore, in addition to reducing defect rates by approximately 2 to 1, the ESD precautions taken earlier probably enhanced the quality of the outgoing HICs as well. This supports the fact that ESD damage can be cumulative and latent.

PWB Assembly Shop — Failure Analysis

As a separate investigation to further understand HIC "A," 302 defective HICs were randomly selected from the PWB assembly shops for FMA purposes. The results, as shown in Figure 4-7, revealed that 39 percent of the defects were ESD–induced with characteristics nearly identical to Figure 4-5.

Additionally, it is possible that a portion of the 48 percent with no trouble found had at one time intermittent defects due to ESD. This is because the device in question has a tendency to latch with subthreshold ESD events and then recover when removed from bias.

In conclusion, it was determined experimentally that ESD precautions taken in the manufacture of HIC "A" resulted in a 2 to 1 improvement at both test operations and, at the same time, substantially improved the outgoing quality. Additionally, a major portion (39 percent) of the defects created during PWB assembly manufacture were ESD induced. If these observations are valid, the general application of ESD precautions would generate manufacturing data to substantiate them. This assumption was confirmed with a financial analysis of the manufacturing process as described in the subsections that follow.

No trouble found	48%
ESD-induced defects	39%
Miscellaneous defects	13%

(Sample Size: 302 HICs)

Figure 4-7. PWB assembly shop failure mode analysis results

Manufacturing Evidence

Comparisons are made in the following subsections between the earlier experimental data and manufacturing data gathered subsequent to the systematic introduction of ESD precautions throughout the shops. The actual precautions taken in each area will be defined. In addition, in 1981, HIC "A" was redesigned to provide better ESD immunity, recoded, and phased-in over a 5-month period. The ESD threshold of this HIC was improved from 700 to 2100 volts.

Manufacturing evidence by itself can only be considered as strong circumstantial evidence. This is because operations of these types are governed by countless variables. However, in conjunction with experimental data, definitive conclusions may properly be made.

HIC Shop Manufacturing Data

Two accounting methods, cost reduction and nonconformance cost, were selected to illustrate the economic benefits experienced subsequent to the introduction of the general use of ESD precautions for all products in the HIC shop. They are each discussed in the following paragraphs.

The first accounting method, cost reduction, is a formal process that requires certification from the accounting organization to validate the calculations and data documentation. In this instance and for all product lines in the HIC shop, a three-month average yield was recorded prior to incorporating ESD precautions. The general use of wrist straps and static dissipative tote trays was then introduced during a three-month grace period. Following this, another set of three-month average yields was established for comparison purposes.

The composite results are shown in Figure 4-8, where the actual yield improvements were 1.3 and 3.3 times better than experimentally predicted (Experiment 2) for the in-process and final tests, respectively.

	Experimental	**Actual**	**Ratio**
First test	2.4%	3%	1.3
Final test	1.2%	6%	3.3

Rate of return on investment — 950%

Figure 4-8. A comparison of actual HIC shop cost reduction results to experimental defect levels show reasonable correlation.

The discrepancy is not particularly surprising since only one of the many products in production was discussed in Experiment 2. This data represents all products in production at the time, and it therefore follows that HIC "A" was not a worst-case situation. In addition, the rate of ***return on investment was 950 percent***; therefore, the prevention techniques used were extremely cost-effective.

The second accounting method, nonconformance cost, is maintained as an independent measurement of manufacturing performance. It is an indicator of both the material and labor financial losses associated with a discarded product. It logically follows that the significant yield improvements described above must translate into major reductions in the nonconformance costs.

Figure 4-9 is a graph of the nonconformance cost performance of the HIC shop during and after the introduction of ESD precautions. It is a plot of the ratio of actual expense dollars to forecasted expense dollars versus time ($/$ versus time). Without specifying the details of the operation, the manufacturing conditions were such that the forecasted expense dollars may be considered essentially constant for the 3-year period shown.

Therefore, the changes in the graph correlate directly to changes in actual expense dollars incurred due to a nonconforming product. Clearly, there is a marked improvement during the course of 1981. In fact, a 20 percent reduction is noted when comparing the first quarter of 1981 to that of 1982.

The introduction of ESD precautions began in October 1980. By the end of the first quarter of 1981, approximately 70 percent of the work positions were equipped with wrist straps. ***The introduction of wrist straps coincides with the sharp decrease in nonconformance cost***.

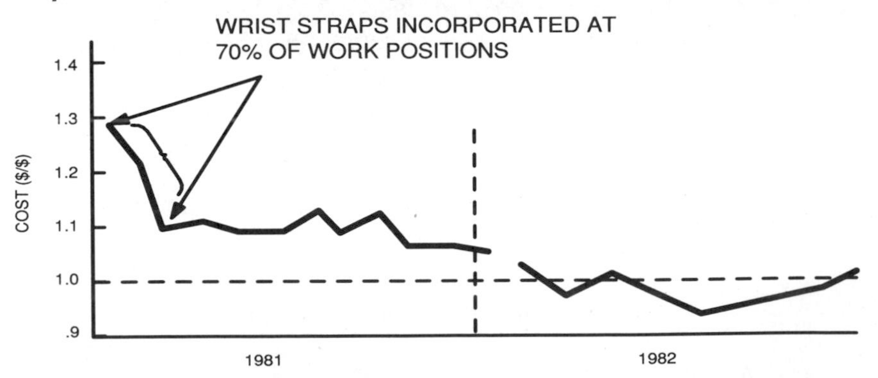

Figure 4-9. HIC shop nonconformance cost ($/$) (actual $/forecasted $)

Additional precautions were phased-in during the balance of the year, which contributed to the continuing downward trend of the graph. These additions included the remaining wrist strap installations, dissipative tote trays and employee training.

In conclusion, the two independent accounting methods presented established strong circumstantial evidence that ESD precautions were extremely cost-effective when applied to the entire HIC shop manufacturing process and that the improvements were consistent with the experimental predictions for HICs. This does not address the impact on PWB assemblies; therefore, the investigation was further extended.

PWB Assembly Shop Manufacturing Data

In February 1982, the PWB assembly manufacturing shop was fully equipped with wrist straps, heelstraps, dissipative table mats, and floor mats, where appropriate. A high degree of employee compliance with ESD procedures was achieved almost immediately as a result of the experience gained in the HIC shop. The impact of these precautions on HIC "A" can be estimated from the experimental data.

Experimentally, it was concluded that at least 39 percent of the HIC "A" defects found in the PWB assembly shop were ESD induced and that precautions taken in the HIC shop had improved the outgoing HIC quality. However, the experimental data had not yet been substantiated with manufacturing results.

The PWB assembly manufacturing results obtained subsequent to the introduction of ESD precautions augmented the experimental results and were documented in four different ways. The first observation was an abrupt reduction of 60 percent in the defect rate of HICs. Next, there was a sharp reduction in the nonconformance cost. Then, a formal cost-reduction investigation established a cost-effective rate of ***return on investment of 185 percent***. The fourth measure of improvement confirmed the enhancement of outgoing quality in a report by the Quality Assurance (QA) Department.

The Quality Assurance Department statistically samples and tests outgoing PWB assemblies for compliance with system requirements during a 24-hour system test. Defects are then analyzed and reports generated for analysis and corrective action, if necessary. One such report separates failures attributable to HICs and is graphically represented in Figure 4-10 for 1980, 1981, and 1982.

HIC failures are plotted as a percentage of the total HICs sampled on PWB assemblies versus time. Two marked improvements (2.1 and 1.6 to 1) are noted, which coincide with the introduction of ESD precautions

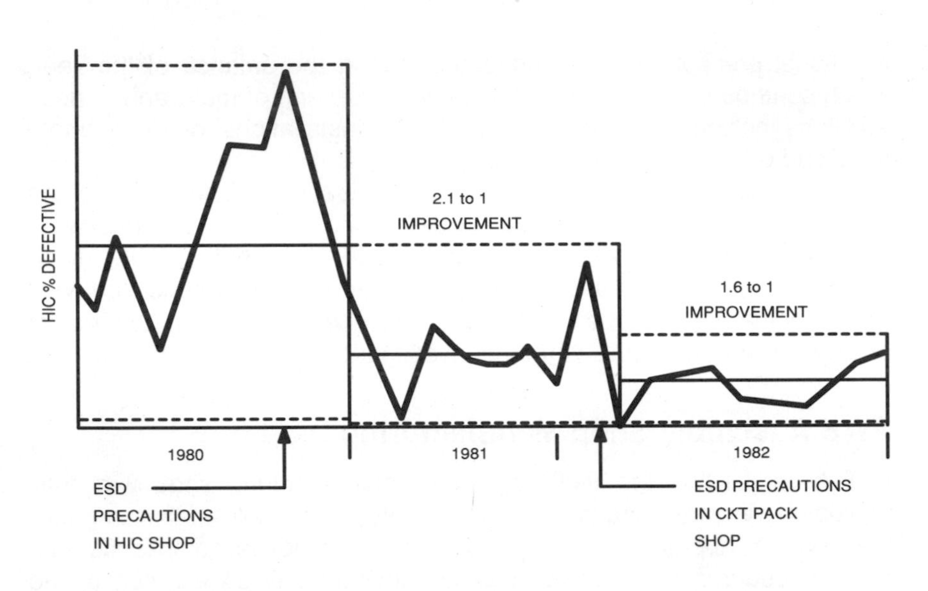

Figure 4-10. Quality assurance sampling of outgoing PWB assemblies—percent defective HICs versus time

first in the HIC shop in late 1980 and, later, in the PWB assembly shop in early 1982. These improvements are consistent with the results of Experiment 2, where outgoing HIC quality was enhanced by 5.5 to 1.

If these observations are valid, the field performance of PWB assemblies produced in 1982 would be better than those produced prior to 1981. This assumption was confirmed by reviewing field return data on a representative sample of PWB assembly codes at comparable points in the product maturity cycle. The failure-in-time (FIT) rates of 1982-vintage PWB assemblies were found to be significantly better than those manufactured prior to 1981.

A composite reduction in QA defect rates of approximately 3 to 1 occurred on two separate occasions, coinciding with the introduction of ESD precautions. Also, a significant improvement in field performance was consistent with the improvements in the outgoing PWB assembly quality reflected in the QA data. At the same time, yield, cost reduction, and nonconformance cost indicators showed marked improvements.

In conclusion, it was predicted, and proven through experiments, that ESD precautions would significantly improve yields in both areas of manufacture, as well as enhance the outgoing HIC quality. Coincident with the introduction of ESD precautions and an ESD redesign, ***four separate measures of manufacturing performance confirmed that the improvements did occur***.

This combination of experimental and manufacturing evidence, in conjunction with the timing, can lead to only one conclusion. ESD precautions, both in design and handling, are not only cost-effective but are also necessary to achieve the highest levels of outgoing quality and reliability. This conclusion applies to both component and PWB assembly manufacture. However, it is recognized that ESD control is not the only contributing factor here. For instance, during the same 3-year period, design, test, and manufacturing improvements were also incorporated. One of the design changes was the HIC "A" redesign which provided better ESD immunity.

Case Study 4: Latent Failure Due to Prior ESD Damage

This case history was selected to illustrate a latent failure due to prior ESD damage in a HIC design using a bipolar silicon integrated circuit.

Experimental Evidence

The first evidence of a problem appeared in the early stages of initial production during a quality assurance sampling when 3 out of 15 PWB assemblies failed the system test. These PWB assemblies had just passed an identical system test as part of the manufacturing process, which did not include ESD protection. Subsequent defect analysis revealed a bipolar junction on HIC "B" with excessive leakage in all three PWB assemblies. Also, the failing external HIC pin was routed directly to another HIC on the PWB and not to an external PWB assembly connector pin.

Later, an unrelated laboratory evaluation of 24 of the same type of PWB assemblies was initiated. The PWB assemblies were put into an operating system, tested successfully, and then left functioning in a secured area. During the next 5 days, 5 of the 24 PWB assemblies failed with the leakage condition described above. Figure 4-11 is a scanning electronic microscope (SEM) photograph (at 4800X) of the junction damage exhibited by all five failures.

Although it is difficult to see, there is a faint trace between the two conductors, indicated by the arrow. The damage was subsequently duplicated by exposure to ESD. The threshold of damage was established at 450 volts HBM for HIC "B" and at 1000 volts HBM for the completed PWB assembly.

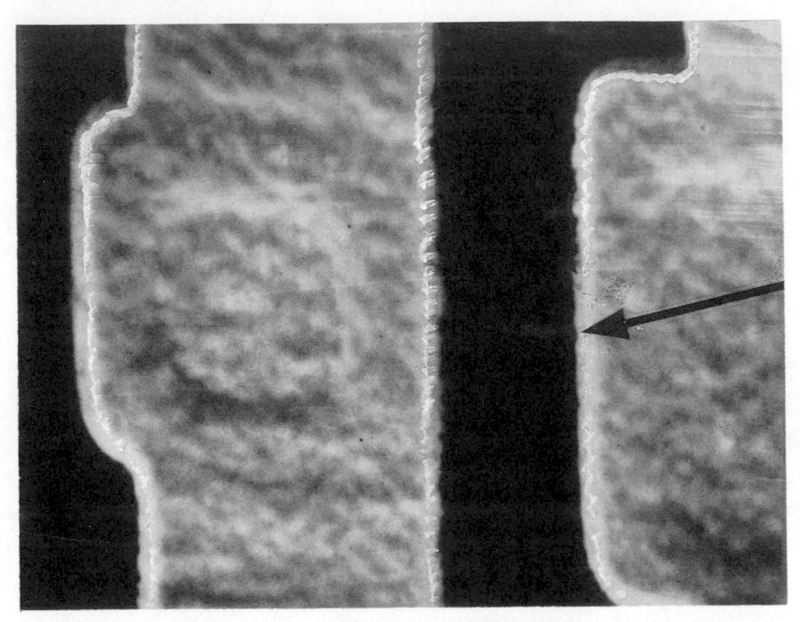

Figure 4-11. Latent ESD failure—bipolar junction

The circumstances surrounding these five failures were such that no one could have touched them once they were operating in the system. Additionally, the testing was done by remote access. Therefore, it is likely that these failures were latent due to prior ESD damage.

At approximately the same time, one customer reported that 17 PWB assemblies out of 31 failed two weeks after being successfully put into service. All exhibited the same leakage condition as the five laboratory failures and were suspected of having latent ESD failures.

In comparing this failure activity to the in-house data above, a statistically significant difference is noted with a confidence level of 99 percent. Likewise, a review of the field data indicated that this situation was extremely abnormal. Therefore, unique and severe conditions triggered the 17 failures. Further evaluation revealed that these PWB assemblies had been expedited through unusual channels in the dry winter months and that they had been transported in EPS trays.

Furthermore, these PWB assemblies were daughter boards and required assembly to the mother board on customer premises. During assembly, it is particularly convenient and almost necessary for the installer to directly contact the conductor on the PWB leading to the indicated HIC pin, thereby increasing the probability of ESD damage. Therefore, most likely the EPS packaging, in conjunction with the circumstances in the field, was probably a major factor leading to latent failure. However, prior damage in the factory or in transit could not be ruled out.

Compared to the number of failures during early production, these failures were insignificant and were the only ones reported. However, on the premise that it was an early warning, response was prompt—ESD precautions were incorporated throughout the manufacturing and shipping process and a Zener diode was added to the PWB assembly to shunt ESD transients to ground. Adding the diode improved the PWD assembly threshold from 1000 volts to something in excess of 15,000 volts. As a longer-term solution, HIC "B" was redesigned to incorporate additional protection.

In conclusion, latent failure due to prior ESD damage was witnessed under laboratory conditions and was suspected of having occurred on customer premises while the PWB assemblies were in service as a result of EPS packaging. This, in conjunction with other reports of latency,[36] supports previous conclusions that ESD damage can adversely affect the reliability of bipolar devices.

Case Study 5: Ultrasensitive Devices

The trend toward including ultrasensitive devices in the manufacturing process calls for a separate discussion of the problems and difficulties that can arise in handling these devices. One such case was revealed with the introduction of an N-type metal oxide semiconductor (NMOS) device that had an ***ESD threshold of 20 volts***. Major problems were encountered during device fabrication as well as during the assembly of PWB assemblies.

Device Testing

Since production levels of this particular NMOS device were still quite low, each device was precious and expensive. This low level of production was due in part to difficulties experienced during the fabrication and testing of this extremely sensitive device. Yields at the device plant were vacillating widely. For instance, during one test operation, an exceptionally high dropout rate was noted. As many as 40 to 60 percent of the devices coming though the line were lost.

It was recognized that the test fixture was made of TEFLON* material and highly prone to charging. Also, only one of several employees was experiencing this high dropout rate.

The employee experiencing the high dropout rate had exceptionally long fingernails. Other employees with shorter nails were not experiencing the high failure rates because they inadvertently touched most of the leads, creating a shunt and grounding the device. Thus, when the device was brought into the field created by the TEFLON material, charge separation was not taking place and the device was not being damaged. On the other hand, the employee who held each device with long fingernails was not grounding the devices on which she worked. Consequently, as these devices were brought into the field, charge separation was taking place. As the test set engaged the device's conductors, discharge occurred causing damage and a high failure rate.

Two solutions presented themselves. The first was to exclusively use employees with short fingernails. The second and more effective, was to redesign the test fixture in question in order to eliminate the TEFLON material and thus the static field. In this way, charge separation could not take place regardless of fingernail length nor could the devices be damaged. Conventional ESD control techniques such as wrist straps did not solve this problem.

PWB Assembly

Extreme fluctuations in PWB assembly yields (Figure 4-12) were occurring during the start of "ramp up," that period during which production quantity begins to increase rapidly in order to meet the ultimate levels of production. Between the months of June and September of 1987, the removal rate varied dramatically between 10 and 30 percent. In actual lot-to-lot observation, ***some lots showed a 100 percent dropout*** in which every single device was defective.

Due to the scarcity of these 20-volt NMOS devices, the cost implications of their continued failure were very high. Therefore, a detailed investigation was undertaken. Through failure analysis, it was determined that the devices were failing due to ESD. In fact, it was demonstrated through failure mode analysis at Bell Laboratories, that

* Registered Trademark of E. I. Du Pont De Nemours & Co., Inc.

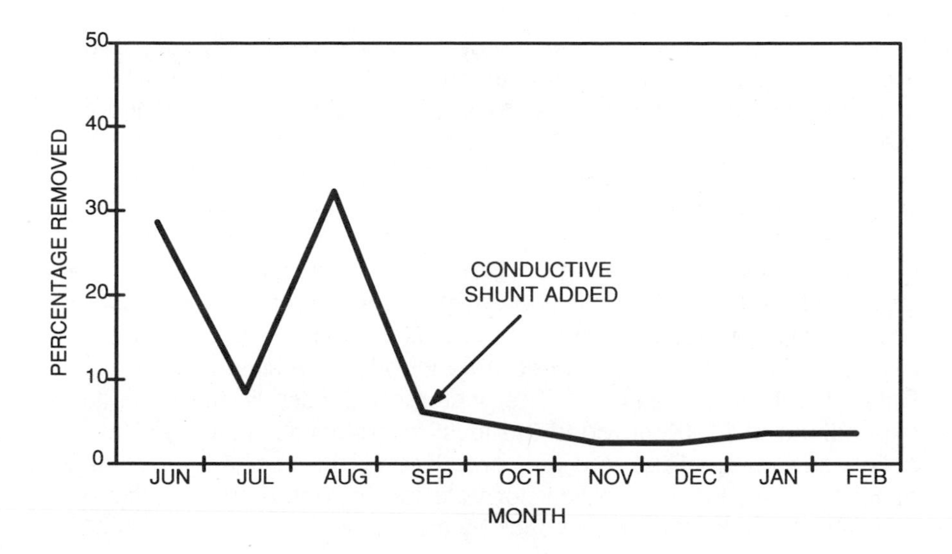

Figure 4-12. PWB assembly yield variation due to ESD damage of an NMOS device with a 20-volt threshold

virtually all of the failures were ESD induced. However, no solution for handling a device that failed at 20 volts was readily apparent.

A special detailed audit was conducted, and a number of people experienced in different aspects of the issue were consulted. A detailed inspection of the manufacturing line was begun, and an action plan of things needing to be corrected was compiled. Based on that action plan, a task force was assembled and assigned to correct deficiencies in the line and to report weekly on what corrective measures had been taken. Because of the extreme seriousness of this situation, the weekly reports were channeled to high-level executives in the company.

Initially, all kinds of extraordinary handling precautions were instituted. Yet, even with all of this special attention and with the fullest compliance with the procedures defined by Class I rating (Chapter 7, "Realistic Requirements"), yields continued to fluctuate dramatically from June through September (Figure 4-12).

The solution to this particular problem was found in the introduction of a "top hat." A top hat is a conductive shunt that is placed on top of a device after it has been assembled to the PWB. As soon as these problem-causing ultrasensitive devices arrived at the assembling operation and immediately after they were assembled to the PWB assembly, the shunt, electrically shorting the leads together, was placed on top. The board was then allowed to go through the production line in normal sequence. The results of the inauguration of that procedure during the month of September are clearly and dramatically recorded in Figure 4-12. By mid-November the removal rate had dropped even further to around 2 percent. In short, by the simple addition of a shunt to the devices, a dropout rate of 30 percent was reduced to 2 percent.

The simplicity of this solution is particularly striking in contrast to more common procedures involving every kind of ESD protective device known to science. The use of so many kinds of precautions eventually becomes difficult to manage. In cases such as this, when an ultrasensitive device is so easily damaged, the extraordinary measure of using a multitude of standard precautions may prove futile as well as expensive. The solution described here introduces a simple shunt into a Class I set of procedures. The incremental cost is trivial. A total expenditure of $1000 provided the level of protection required. Yet the dollar savings realized on the production line excluding overhead expenses, reached $6.2 million per year for this one device on this one line. That is an impressive payback by any measure.

One additional benefit derived from this case was the impact that it had on the design community. Asked to justify a threshold of 20 volts for the NMOS device involved in the project, designers responded by redesigning the device and raising the level of sensitivity to 1000 volts HBM, a remarkable accomplishment.

In conclusion, this case study makes it clear that ultrasensitive devices can present a potential threat to production lines that may result in lost production and lost sales. The financial implications are particularly unattractive when the cost of lost sales is added to the cost of lost materials. Note that the PWB assembly, in its final configuration, is enclosed in a metal housing. Consequently, this ultrasensitive device has always been well protected in the field and has a low return level for ESD defects.

As a direct result of the experience outlined in this case study, minimum design requirements were modified and a new set of handling requirements (Class 0)[11] were established and added to the AT&T handbook, as detailed in Chapter 7, "Realistic Requirements." It was apparent that ***a cookbook approach to establishing handling criteria for ultrasensitive devices would not work***. For example, it is likely that some of the automated equipment used in the assembly process was causing the problem solved by the application of a top hat. Clearly, all of the wrist straps and ionization units in the world would not have solved this problem. Adding a shunt was not only necessary but sufficient to protect the device at great economic benefit. The solution offered tremendous economic leverage. In addition, a Class 1 shop was allowed, through this solution, to continue to do business as usual while protecting an ultrasensitive device. Training considerations were minimized, and the impact on personnel significantly simplified.

Conclusion

The experimental approach presented in this chapter has proven to be an effective tool to assess, in advance, the benefits of introducing ESD precautions into a manufacturing process. The experiments, with a reasonable degree of accuracy, predicted significant yield and outgoing quality improvement for both components and PWB assemblies. The design, execution, and failure mode analysis of such experiments ultimately determine whether or not a definitive conclusion can be reached. For instance, selecting an ESD-sensitive test vehicle increases the likelihood of detecting a difference between two populations. These experiments, although time consuming, are relatively inexpensive to run and provide ample justification for equipping a product line and ultimately a factory with ESD precautions.

However, it is the combination of experimental and manufacturing data that provides compelling justification for providing ESD protection systematically throughout manufacturing facilities. The evidence, as presented, establishes that ESD precautions are cost-effective for the protection of even older bipolar designs and contribute substantially to overall improvements in outgoing product quality and reliability. It can also be assumed that as trends in technology continue to produce devices of ever-increasing susceptibility to ESD, the economic benefits of incorporating ESD precautions will increase accordingly.

Therefore, based on the cases presented here, it is recommended that a systematic ESD prevention plan that extends from design through customer acceptance be implemented. The design effort is logically directed at the more sensitive devices, while handling precautions are judiciously applied to all solid-state devices such that the more susceptible are provided increased ESD protection.

For facilities faced with having to justify implementation of a new program of ESD controls, the case histories reviewed in this chapter provide irrefutable evidence of the economic benefit to be gained by such a program. Some returns on investment recorded here are as high as 1000 percent. In the example of the NMOS device in Case Study 5, ***an investment of only $1000 resulted in a savings of $6 million***. While this is an isolated incident, it is patently clear that an outstanding level of return can be expected from any judicious investment in ESD precautions.

ESD precautions produce yield improvement across the board. All product lines gain from ESD control, although the extent of that improvement will vary among product lines. With the introduction of an ESD program, it is reasonable to expect an average yield improvement of approximately 5 percent, as well as reliability and quality improvements of the outgoing product.

It is also likely that certain catastrophic situations will occur, resulting in very poor yields that could approach zero, such as that described in the NMOS case history. At times, variations in yield can appear so extreme as to suggest multiple causes and to require prompt action. The ideal resolution is to solve such problems on the spot. For that, a local expert is required. There is a need for someone close by who can solve problems of that complexity in a timely, economically sound manner.

It is even more important to be able to prevent damaging situations. The design of an ESD control plan should take into account the probability of such occurrences so that they can be systematically anticipated and prevented. The resulting improvement in yields can be every bit as dramatic as some of those illustrated here.

In many cases designers hold the key to ESD success. They must understand ESD phenomenon and its impact and how to design-in the highest level of protection from ESD damage. See Chapter 5 "Designed-in Protection and Product Testing" for more details.

Points To Remember

- The type of scientific analysis presented in this chapter provides a solid basis for realistic control measures.

- The combination of experimental and manufacturing data provides compelling justification for ESD control measures.

- A strong management commitment to a corporate ESD control program can be enlisted by using the data provided in this chapter or by conducting similar experiments to establish the economic significance of ESD control.

- A review of papers presented at EOS/ESD symposia will yield additional convincing evidence of the economic benefits of ESD control.

- ESD controls can lower operating costs with a rate of return on investment of up to 1000 percent and enhance outgoing quality and reliability.

- Two independent accounting methods confirmed that ESD precautions were extremely cost-effective in the manufacture of HICs.

- Quality assurance sampling of outgoing PWB assemblies revealed a 3 to 1 improvement in defect levels, which coincided with the introduction of ESD controls in the manufacturing process.

- As technology trends continue to produce devices of ever-increasing susceptibility to ESD, the economic benefits of incorporating ESD precautions will increase accordingly.

- ESD precautions, both in design and handling, are not only cost effective but are also necessary to achieve the highest levels of outgoing quality and reliability.

- To avoid excessive implementation costs, ultrasensitive devices require extraordinary measures that must be carefully tailored to the specific situation.

Chapter 5

Designed-In Protection and Product Testing

Considerable effort is required to minimize ESD effects in manufacturing. However, the control program described in this book would be largely ineffective without some protection built into the devices. Manufacturing using metal oxide semiconductor (MOS) devices without specific protection circuitry would be extremely expensive if not impossible. Furthermore, once the devices leave the factory in circuit assemblies, they again become vulnerable during installation, maintenance, and repair. These functions are often performed by someone other than the equipment manufacturer. Thus, protection is required to survive this more hostile environment. Since performance requirements may limit the on-chip protection of some devices, additional protection may be required at the PWB assembly or system level. Special design techniques are also required to protect operating equipment from experiencing ESD-induced upset or soft errors as well as errors in data transmission.

Device Protection

To protect devices from ESD, special ESD protection circuits are placed adjacent to each pin on board an integrated circuit. These circuits absorb the discharge without damaging either the protected circuit or themselves. The types of protection used depend on the device technology and fabrication process. Here-in we will describe a protection scheme used in the AT&T Microelectronics complimentary metal oxide semiconductor (CMOS) devices. Descriptions of other protection schemes are given in the References.[32,37-40]

CMOS is the technology of choice for most very large scale integration (VLSI) devices today. The reason for this is that the power consumed by CMOS circuits is relatively low when compared to bipolar or NMOS (n-channel MOS) devices. The primary failure modes that are observed when CMOS components are "zapped" by ESD are the breakdown of the thin oxide in the MOS structure, leaky p-n junctions, dielectric breakdown, and fusing of conductors. As VLSI devices are scaled down, the gate oxide becomes thinner (150 μ in state-of-the-art devices) with lower breakdown voltages, junctions become shallower (more prone to degradation), dielectrics become thinner, and conductor lines become narrower (more fusing susceptible). Thus protecting these circuits against increasingly difficult.

A typical CMOS input protection circuit is shown in Figure 5-1. The primary protection devices are the dual diodes that are connected directly to the bond pad. The diode to Vss provides the discharge path for negative pulses, and the diode to Vdd is the device that conducts during positive pulses. The resistor along with the gate capacitance provides an RC time delay that slows the charging of the gate. The additional dual diodes provide discharge means to Vdd and Vss of the carriers that have managed to go through the resistor/distributed diode. The layout of the input protection is shown in Figure 5-2. The primary diodes are laid out around the bond pad to evenly distribute the current and to minimize the area consumed by the protection device.

CMOS outputs are protected by their parasitic drain-to-well diodes associated with the CMOS process: the n-channel pull-down device contains the n+/p well diode, and the p-channel pull-up inherently includes the p+/n well diode as shown in Figure 5-3. The layout of the output protection is shown in Figure 5-4. Since these diodes are not optimized to conduct large currents, especially for slow, small drivers, an additional set of diodes is placed around the bond pad.

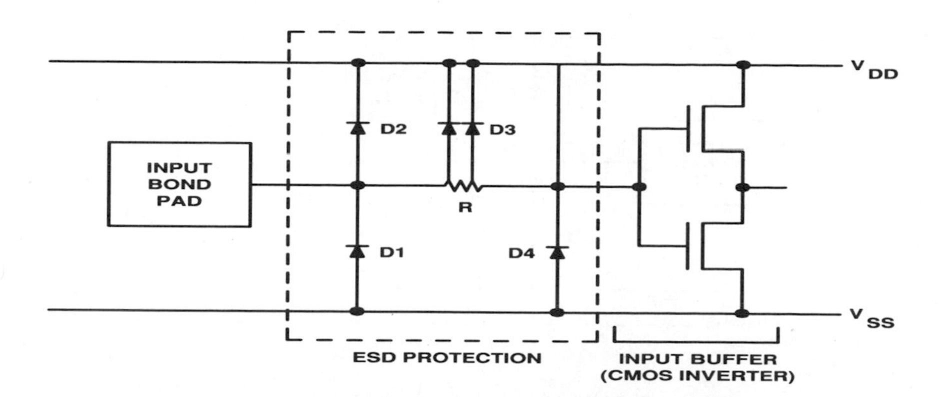

Figure 5-1. Typical CMOS input protection circuit

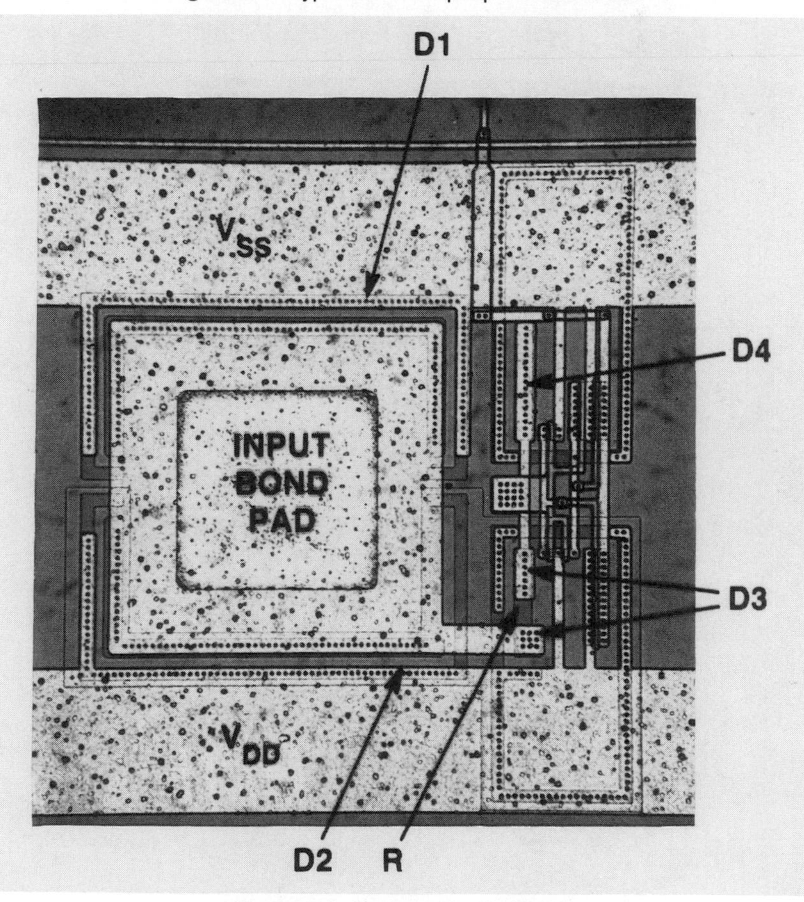

Figure 5-2. Input protection layout

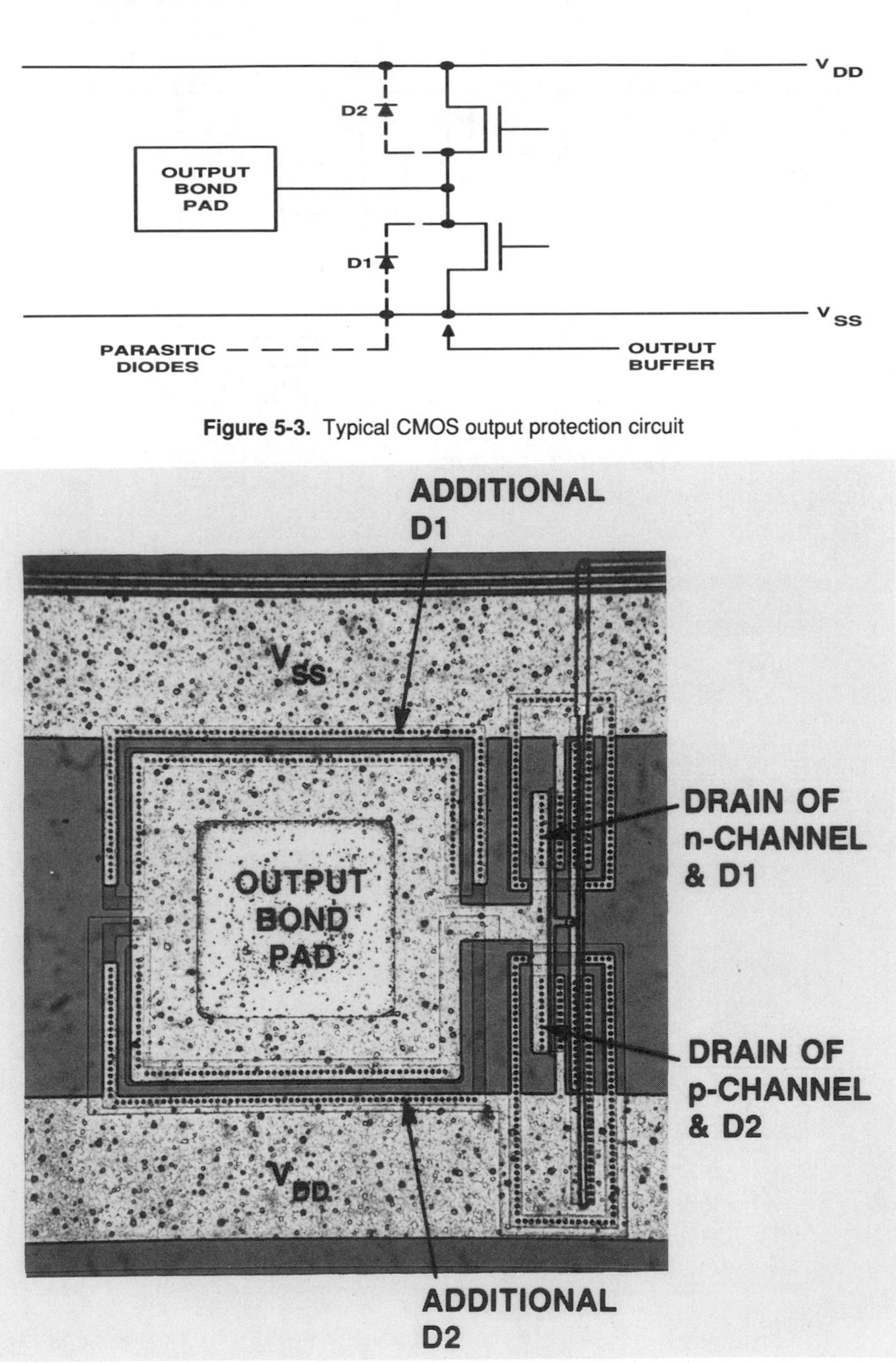

Figure 5-3. Typical CMOS output protection circuit

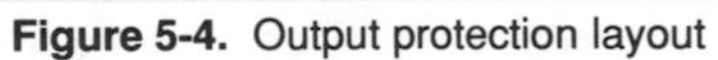

Figure 5-4. Output protection layout

Recently, with the advanced CMOS processes that use lightly doped drains (LDD) and silicide source/drain, output buffers are becoming more sensitive to ESD.[41-43] Failures have been found in the n-channel driver and are either caused by thin oxide damage or by degradation of the n+/p well junction. The damage or degradation has been caused by an unevenly distributed current during the snapback of the parasitic npn device that is associated with the n-channel MOS field effect transistor (MOSFET). To help protect the output buffer, a similar lateral npn device that does not have a thin oxide is used. This npn (or snapback) device is placed between the bond pad and Vss.

Protection at the PWB Assembly Level

Two very different issues must be considered for proper ESD design at the PWB assembly level. First, the PWB assembly and its devices must be physically protected. Contrary to popular belief, devices do not necessarily become less susceptible to ESD failure once placed on PWB assemblies. In fact, some devices may become even more susceptible to ESD. The ESD protection provided by the devices is designed primarily to withstand manufacturing and handling procedures during their processing and assembly. However, once the devices are assembled on a PWB assembly and shipped, they are no longer under the control of the manufacturer. It is at this time that the devices may be most vulnerable to ESD damage. Field failure data from the regional telephone companies substantiates this fact. ESD has been identified as the cause of up to 20 percent of the PWB assembly failures in some systems. As devices become more ESD sensitive, PWB assembly level protection will become even more necessary. The following are techniques that can be used to reduce the ESD sensitivity of PWB assemblies:

a. Restrict access to connector terminals.

b. Use decoupling capacitors wherever possible.

c. Provide a perimeter guard ring tied to the appropriate ground source.

d. Apply specific shielding and surge protection (diodes) to the most sensitive devices.

e. Restrict access to sensitive devices on the PWB assembly by using coatings or physical shields.

f. When coatings or shields are not possible, restrict access to conductors leading to sensitive device pins by routing the conductor paths on inner layers of the PWB or on the component side of the PWB assembly.

g. Avoid routing the sensitive leads of devices to the board edge, connector, or other physically accessible points.

A final design technique that greatly effects PWB assembly susceptibility is proper component selection. Clearly, components that are the least sensitive (have the highest withstand voltage) to ESD should be used whenever possible. ***Functionally identical components from different suppliers may have very different ESD withstand voltages.*** Therefore, correct component selection is a critical part of proper ESD design at the PWB assembly level.

The second major area concerning proper ESD design at the PWB assembly level involves reducing the effects of high-frequency radiation on the system operation. This radiation is typical when a discharge occurs to metallic housings near PWB assemblies, especially faceplates and latches. Since this is an electromagnetic radiation problem, basic electromagnetic interference (EMI) techniques should be used to minimize these effects. These techniques include:

a. Using multilayer boards whenever possible

b. Tying metal faceplates and latches to the appropriate ground by using a perimeter guard ring

c. Minimizing signal-to-ground and power-to-ground loops by using appropriate routing and power and ground plane/grids.

System Level Protection

Proper system level ESD design should protect the system from physical damage and electromagnetic fields produced by an ESD event. The most effective way to protect a system from both of these effects would be to enclose the system in a Faraday cage (Figure 5-5). However, cost and accessibility for operation typically preclude such complete protection. Therefore, system level design should minimize the ability of an ESD arc or its related electromagnetic field to affect the system.

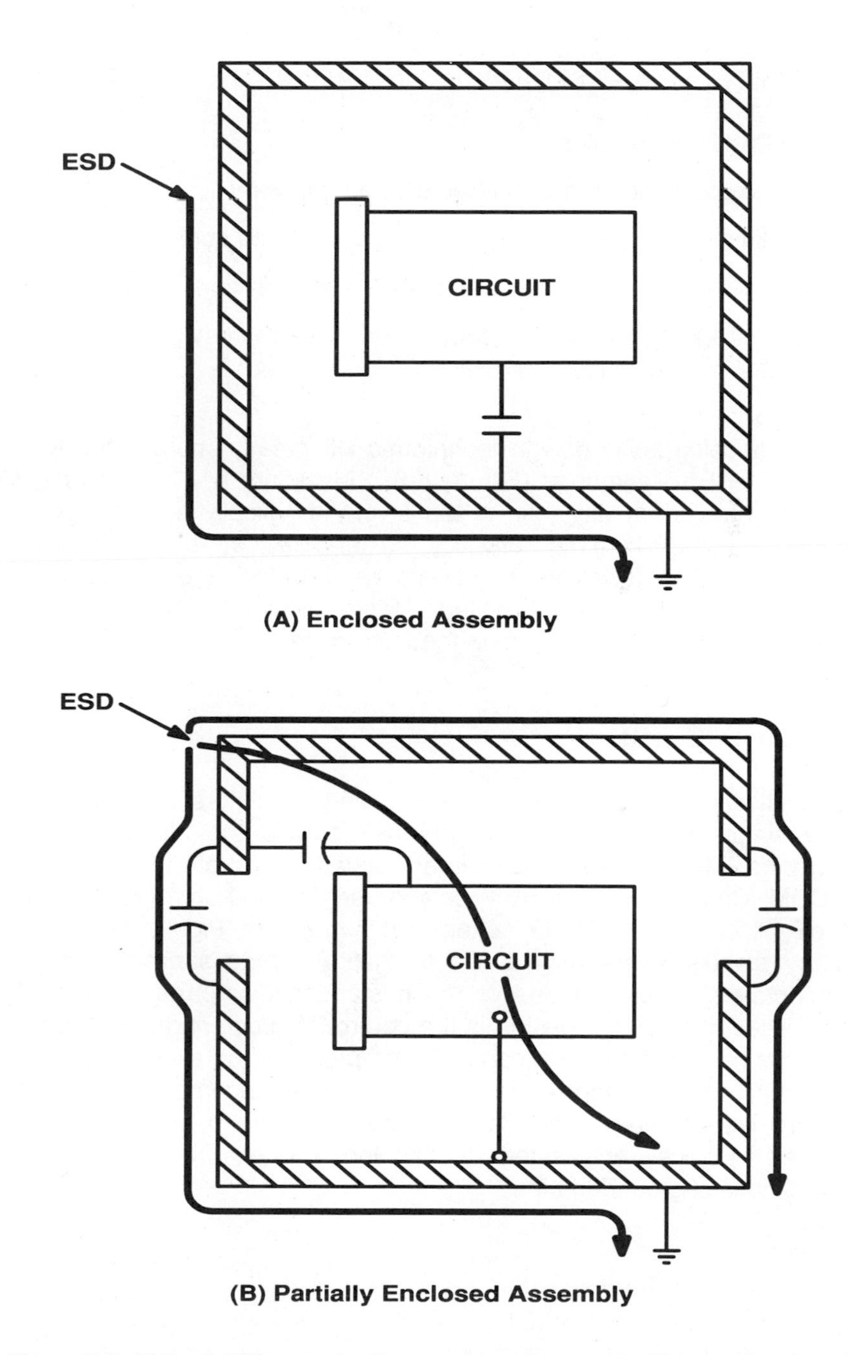

Figure 5-5. Flow of ESD current with an enclosed assembly (Faraday Cage) and a partially enclosed assembly

In general, system level design techniques include:

a. Shielding all cables
b. Using common-mode chokes where necessary
c. Minimizing all physical openings in covers and doors
d. Grounding all exposed components and metallic structures
e. Sealing openings in doors and covers using conductive gasketing or similar hardware.

Properly using these design techniques will greatly improve the total system level ESD immunity (Figure 5-6). Maximum application of the design techniques at any one of the levels is typically not enough to ensure an ESD immune product. Therefore, at each level of development, the designers must ensure that the appropriate ESD protection techniques are applied. More detailed information on system-level design is contained in the References.[44-46]

Device Testing

ESD testing is done using equipment called simulators or "zappers." These simulators produce as nearly as possible the idealized ESD waveforms. Microelectronic devices are tested using both the HBM and the CDM (Chapter 3). The HBM device uses a 1500-ohm resistor and 100-pF capacitor. A simple schematic is given in Figure 5-7. Two industry standards are now available that are quite similar.[47,48] Most manufacturers either use one of these standards or parts thereof. A critical issue in simulator design is the control of circuit parasitics. Early simulators had poor parasitic control[49] and produced unreliable results.[50] More stringent calibration requirements have been incorporated into the standard test methods, and data correlation between simulators has improved. Devices are tested by zapping a single pin with various combinations of grounded pins.

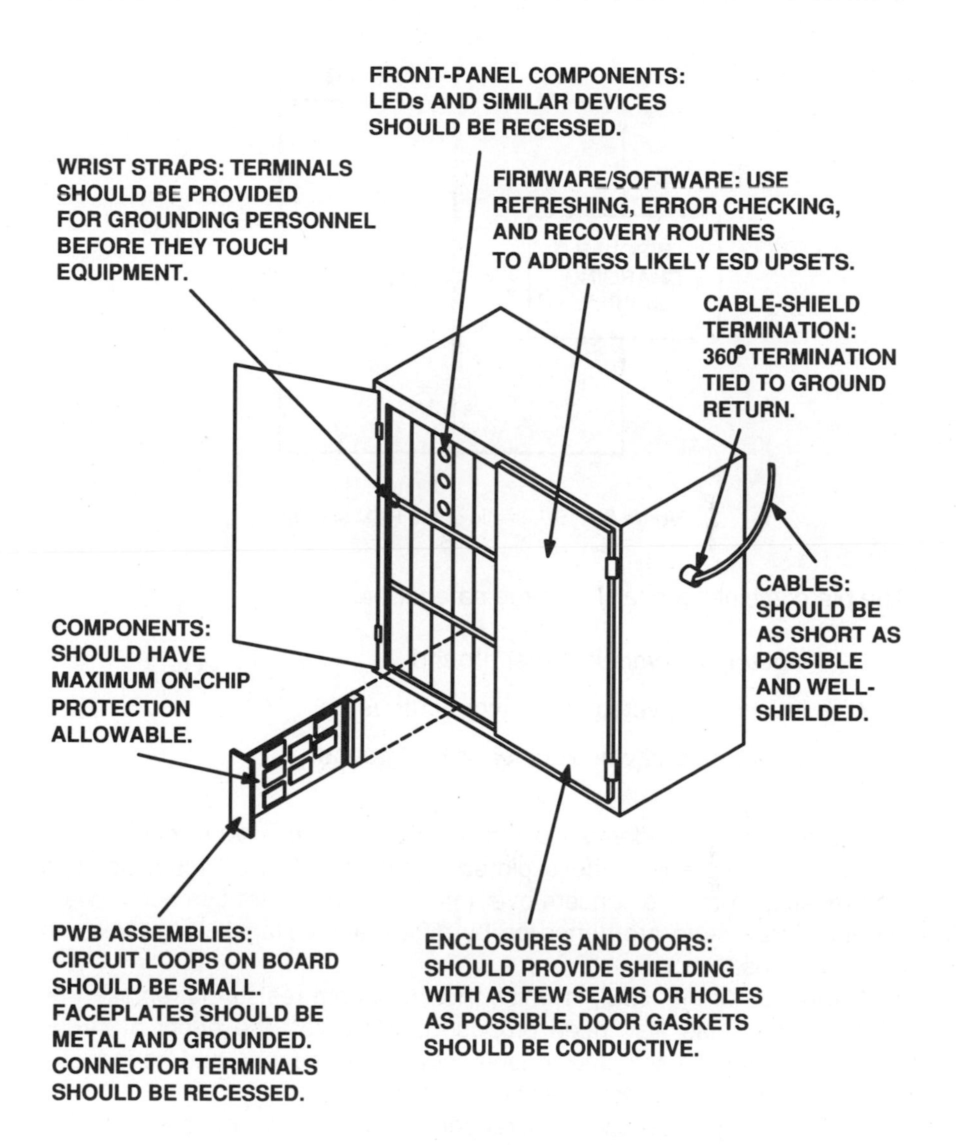

Figure 5-6. Good ESD design requires attention to detail.

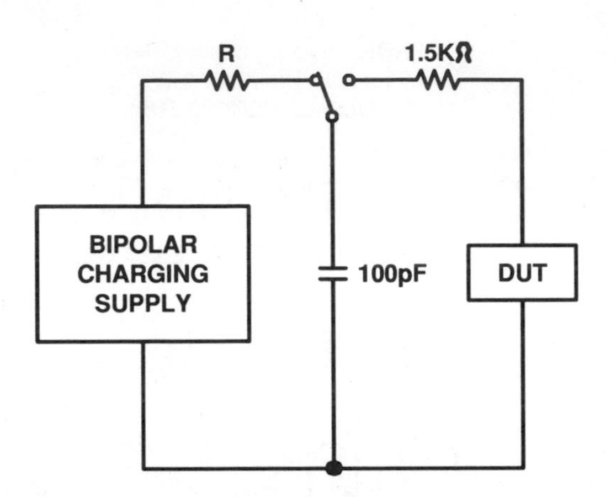

Figure 5-7. Schematic of the HBM simulator

The pin combinations fall into three categories:

1. Input/output versus power/ground
2. Input/output versus other input/output
3. Power/ground versus other power/ground.

There is still considerable controversy over how extensively each of these categories should be explored. Category 3 is a recent addition and reflects a growing concern over internal failures that can occur when parasitic transistors are turned on by differential voltages on power and ground buses.[51]

A variation of the HBM, in which the 1500-ohm resistor is replaced by a dead short, has also been proposed. However, the effect of circuit parasitics in this "machine" model is much more severe, and no industry standard is likely to be forthcoming in the near future.

CDM testing is receiving increased attention. A committee of the EOS/ESD Association is working on a draft test method for the end of 1990. AT&T has been doing CDM testing of devices since 1984. Until recently, this method was limited to testing devices in through-hole mounted dual in-line packages. In this method, the device is directly charged through one of its pins and discharged using a relay (Figure 5-8). A new CDM method based on the principle of static induction[52] (Chapter 3) has been introduced; it can easily test a wide variety of surface-mount packages and solves a number of other technical problems with the old method.

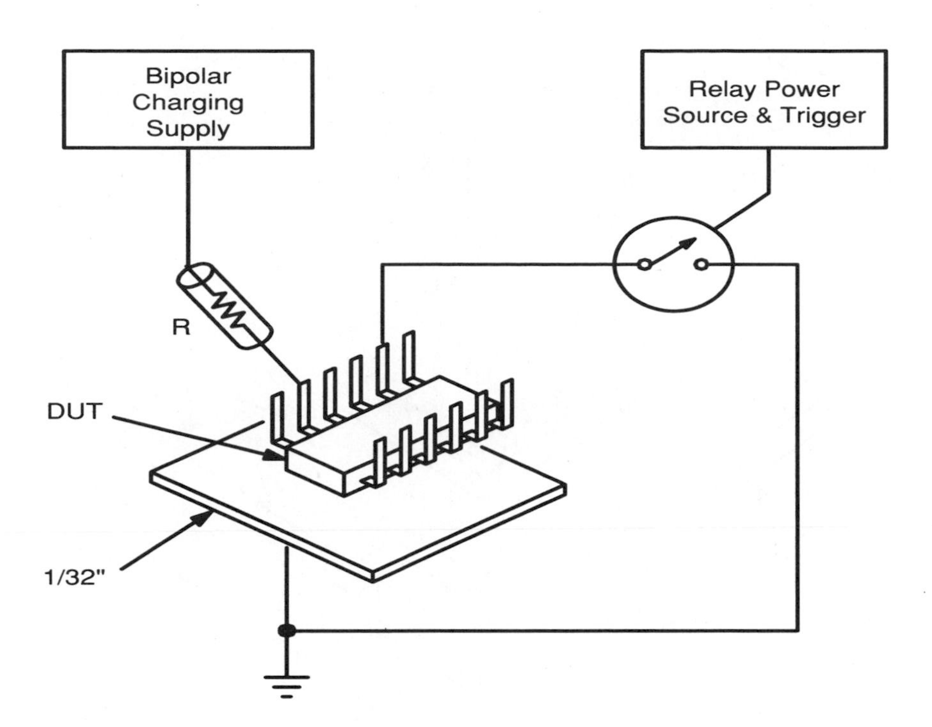

Figure 5-8. Schematic of a CDM simulator using direct device charging and relay discharging

A schematic of this method, called the field-induced CDM (FCDM) is shown in Figure 5-9. Because relays can introduce unwanted parasitic inductance into the discharge path, the FCDM uses air discharge to a grounded pulse. More information on ESD test equipment is given in Chapter 6.

Some important points to remember about ESD testing of devices are:

1. Since ESD sensitivity is extremely dependent on circuit layout, all new designs should be tested.
2. Testing should be done as early in the design cycle as possible so that manufacturing and system level trade-offs can be identified.
3. HBM sensitivities are relatively independent of package type. However, larger packages have lower CDM thresholds (because of the higher capacitance).

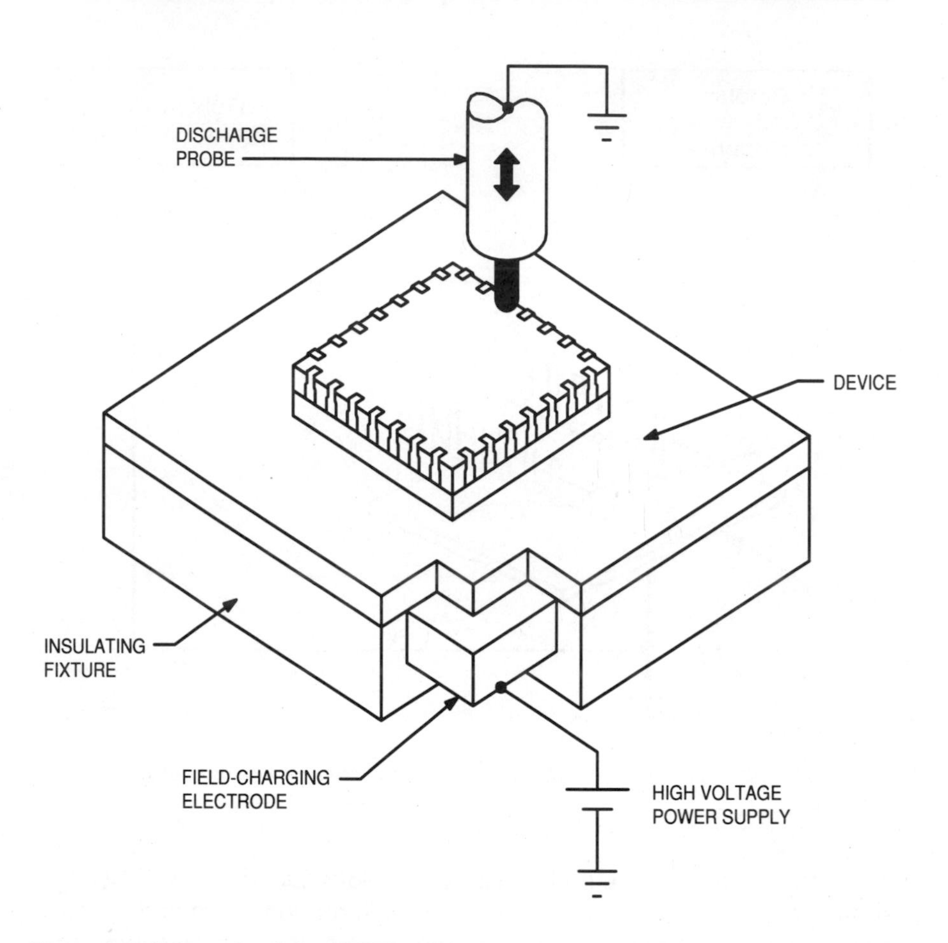

Figure 5-9. Schematic of the FCDM simulator, which uses induction charging and air discharging

4. ESD is a high-frequency phenomenon. Take care to minimize the parasitic effects of sockets or ground returns. That is, don't think DC!

ESD withstand threshold data should be included in device data sheets.

PWB Assembly Testing

At the present time, no agreed-upon method for testing unterminated (that is, not installed) PWB assemblies is available. An industry group has been formed to consider such a test.

System Level Testing

Electronic equipment is tested using hand-held or tripod-mounted ESD "guns" (Figure 5-10). Although there are some exceptions, most tests are based on International Electrotechnical Commission (IEC) specification 801-2 (Table 5-1), which calls for a 150-pF capacitor and 150-ohm resistor. This test also specifies air discharge. That is, the tip of the gun is allowed to reach the specified test voltage and is subsequently moved toward an exposed portion of the equipment under test (EUT). If the electrical field between the tip and the EUT exceeds the breakdown level, an arc discharge occurs and current flows over or through the equipment. An alternate method that can be used is contact discharge, where the discharge occurs in a vacuum relay. This is sometimes referred to as current injection. Air discharge has the advantage of being more realistic, but the test conditions, such as the speed of approach, are difficult to control. Most telecommunications equipment has been tested this way over the past several years, and the major standard for central office equipment has adopted this method. However, the IEC 801-2 document is under revision, and it appears that contact discharge will be preferred in the future. In addition, a standard proposed by the Comite' International Special des Perturbations Radioelectriques (CISPR) specifies contact discharge. The CISPR document will be the basis for the ESD requirements adopted by the European market in 1992. Actual voltage requirements and failure criteria tend to be product specific and negotiated between the manufacturer and user.

ESD Design—Dealing with the Trends

In the future, the drive for higher speed and performance will continue to push ESD thresholds lower. At the same time, the trend towards high-throughput manufacturing processes will continue to transform assembly factories. High-speed transmission equipment will require greater attention to noise reduction, while a competitive marketplace will

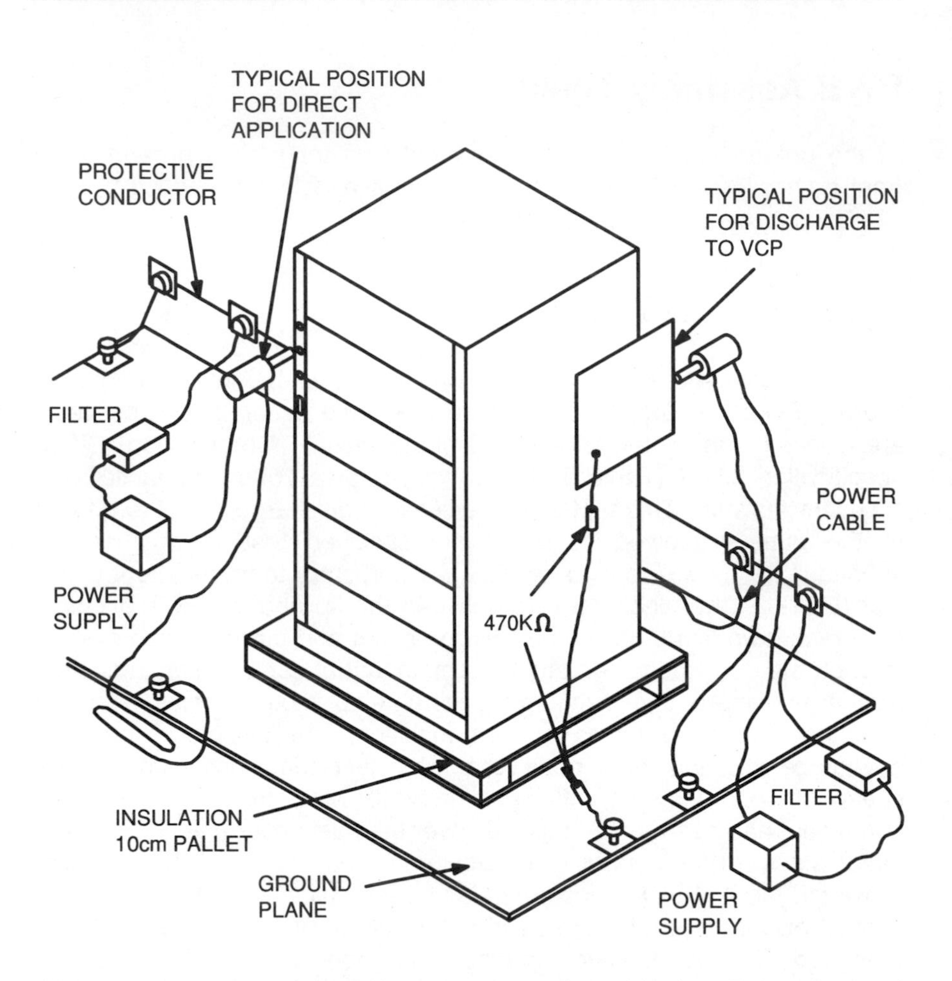

Figure 5-10. Typical set-up for ESD testing of operating equipment. Both a direct and indirect discharge to a vertical conducting plane (VCP) are required.

demand cost reduction wherever possible. At each of the design levels, techniques are available that can be applied. Even devices that operate at more than 1 Gbit/sec can tolerate some protection if the device designer works with the system designer to deal with the extra capacitive loading.

While the trends toward higher scale integration and high speed do present technical challenges, some of the major impediments to producing optimal ESD performance are nontechnical. The spread of thresholds in MOS and bipolar devices indicated in Figure 3-11 is due less to intrinsic device sensitivity than to the amount of effort given to ESD protection by the designer and the amount of emphasis given by

Table 5-1. Summary of Industrial Standards

Standard or Organization	Products	Component values RΩ	Component values C(pF)	Maximum Voltage (kV)[1]	Discharge Types[2]
MIL STD 883	Microelectronics	1500	100	8	dc
EOS/ESD-DS5[3,4]	Microelectronics	1500	100	8	dc
IEC 801-2(1984)	Industrial Eqpt	150	150	15	da,ia
IEC 801-2(1989)[4]	Industrial Eqpt	330	150	6	dc,ic
SAE	Automotive	250	200	15	da
EIA PN-1361	Terminals	150	150	15	da
CISPR (Europe)[4]	Telecomm.[5]	330	150	8	da,ia
		330	150	3	dc,ic
ANSI C63	General	330	150	6	dc,ic
		15	150	6	dc,ic
ANSI (T1Y1)[4]	Central Office	150	150	15[6]	da,ia

Notes:

1. Indicates maximum test voltage, not necessarily requirement.
2. Discharge types are direct to equipment (d) or to external plane (i) air discharge (a) or contact (c).
3. Differs from MIL STD 883C in calibration method and pin combinations.
4. Proposed.
5. Telecommunications.
6. Requirement is 8 kV.

the design organization. In some cases, the designer may be unnecessarily limited by the fabrication process. Thus, ESD performance should be considered during process development—before the first design is completed. Once the process is frozen, the device designer may have little flexibility in providing adequate protection.

As we move into submicron CMOS technologies, providing sufficient CDM protection compatible with high-volume manufacture will be important in keeping manufacturing (assembly) costs down. This is aggravated by the fact that CDM thresholds decrease with increasing package size (capacitance).

In many companies, attention to ESD design tends to occur in bursts every few years. Each burst is usually motivated by some catastrophic event: a major field failure, serious yield losses, or lost equipment sales. Though this burst may last months or years, ESD design eventually gets reduced to selecting from a few buffers in a design library or specifying the same cabinets as the previous designer did without regard to changes in technology or manufacturing requirements. Because ESD failures do not occur every day, at least not visibly, there is a tendency to give ESD a low priority in design reviews. This attitude is reinforced by successful manufacturing programs, such as those at AT&T, which seem to suggest that control of the event is the complete solution. The "burst" mode of determining ESD design priority will not be adequate in the face of technology trends. Providing for a more disciplined and continuous approach to ESD design starts with the design and product managers. These managers must provide adequate resources and plan for testing time. This clearly adds some cost to development, but the return on this investment has been high in the past and will be even higher in the future.

Points To Remember

- Factory controls alone cannot prevent ESD damage. Design-in protection is also necessary.

- Devices, assemblies, or systems with poor designed-in protection may survive in the factory because of the rigorous application of controls but then fail in the relatively less controlled external customer environment.

- Sound ESD protection should be designed-in at the device, PWB assembly and equipment level.

- As device features are reduced in size, the ESD withstand thresholds tend to decrease.

- There are two primary ESD tests required for microelectronic devices: the human-body model and the charged-device model.

- ESD sensitivity is extremely dependent on circuit layout. Device technology (MOS, bipolar) is a relatively weak indicator of protection.

- Testing should be done as early in the design cycle as possible.

- ESD is a major cause of equipment malfunction and transmission errors. Design techniques are available to minimize these effects.

Chapter 6

ESD Test Facilities

Testing is an essential part of any ESD program, as it permits the coordinator to scientifically evaluate many critical aspects of the program. Areas where testing can be useful include:

- Defining procedures and requirements used in an ESD Control Handbook
- Qualifying ESD control products
- Providing purchasing with the necessary information and requirements to buy and monitor incoming items
- Providing information, procedures, and requirements for ESD Auditing
- Solving manufacturing problems
- Providing FMA for devices and systems and
- Testing and qualifying devices and/or systems prior to shipment.

The numbers and types of test equipment needed for an ESD program can vary greatly with the type of program desired. Available test facilities and equipment range from simple field meters to elaborate simulators and FMA instruments. The test equipment can be organized in a variety of ways, ranging from a small cart or suitcase of equipment used in a field audit program to an elaborately equipped laboratory used to qualify devices and/or systems, test materials and to do FMA.

The information that follows highlights a number of available instruments and tells how to use some of the more common ones and how they might be organized for a manufacturing application.

Before purchasing any equipment, it is necessary to evaluate the intended application and decide what equipment will be best suited to that use. The following list of questions regarding options should be useful in making that determination.

1. Is it for portable or fixed plant use?
2. Is AC or battery operation required?
    a. If AC operation is necessary, what are the voltage and current requirements?
    b. If battery operation is required, what type is it? Is it readily accessible and easy to change? Is it rechargeable and does it come with a charger? Does it have a low battery indicator?
3. Is it necessary to ground the instrument for an accurate reading, and are ground leads provided? Will nearby ground planes affect the readings?
4. Are probes required? If so, what types are available? Are the leads compatible with existing equipment?
5. What type of output is available? Does it have a digital, analog, or LED display? Does it require a strip chart recorder or equivalent?
    a. If a record is to be kept, what type of recorder is it compatible with?
    b. If a printer is required, what interface cables and format are needed? Does it come equipped with interface cables?

6. Is it easy to operate?

    a. Are switches and controls convenient and easy to use?

    b. Is thc instruction manual informative and easy to use?

7. Can the unit be calibrated easily, and is it traceable to the National Bureau of Standards?

The following sections will show one means of setting up three types of testing facilities, including a Field Audit Kit, a General Laboratory (with a portable test cart), and an Analytical/FMA Laboratory.

Field Audit Kit

A field audit kit is portable and can be hand carried in a custom carrying case or on a cart. The kit is primarily intended for the plant ESD inspector or auditor and can be easily transported to remote locations either within the building or off premises. The test equipment consists of basic instrumentation used by the inspector to audit control facilities and should include the following:

Cart — With the proper grounding cord and/or conductive wheels and dissipative surfaces—this is used to transport audit equipment.

Carrying Case — Static dissipative foam lined, with an easily partitioned interior to accommodate portable test equipment.

Combination SRM/RTG Meter — Surface Resistivity Meter (SRM)/Resistance to Ground (RTG) Battery operated, with leads, used to measure resistances, both surface and to ground, of many items. Leads need to be shielded.

Wrist Strap Checker — Used to check wrist straps, foot straps, etc., and the resistance to ground of tabletops and table mats.

Footwear Checker	With foot plate—needed only if footwear falls outside of the operating range of a wrist strap checker.
Calibrator	For wrist strap and footwear checkers—to verify proper calibration.
Multimeter	Used to check resistances and voltages.
Ion Current Meter	Used to check the balance and amplitude of the output current of portable ionizers. Also called a residual voltmeter.
Residual Voltmeter	To check the balance voltage of portable ionizers. Should be used in conjunction with an ion current meter since a zero voltage balance could be indicated by a defective ionizer.
Electrostatic Locator	Noncontacting field meter to measure static potentials on objects.
Horsehair Brush	Used to generate charges on objects, by stroking, in order to check triboelectric charging.
Wrist, heel, and toe strap	Used by auditor both for proper grounding and for demonstrating proper use and test methods.
ZEROSTAT* pistol	A commercially available (from many record stores) unit used to neutralize charges. Looks like a small red pistol. Consists of a piezoelectric generator that generates negative charges when the trigger is pulled and positive charges when released.
Documentation	Inspector/Audit Manual, ESD Handbook, ESD Bulletins, sample forms, etc.
Reject Stickers	Used to tag items found to be in violation.
Megger Meter	A portable meter (100 volts) that enables the measurement of high resistances with two NFPA probes.

* Registered trademark of Discwasher, Inc.

Two NFPA Probes Two 1/2" diameter, 5-lb probes with compliant conductive surfaces. Used with a Megger meter to measure surface point-to-point resistance (and resistances to ground) using the procedures outlined in NFPA (National Fire Protection Association) Standard 99-89.

Miscellaneous Scissors, tape, batteries, signs, ground cords and other leads, isopropyl alcohol, wipes, ruler, multiple outlet extension cord, and tribocharging materials such as untreated plastics.

General Laboratory

The General Laboratory should be the base of operations for ESD work done within a plant. Equipment must be versatile and usable by technicians as well as the ESD coordinator. In general, the equipment is more sophisticated than that needed for the field audit. Some of the equipment is portable, and other items are stationary. Many of these items are described later under Instrument Descriptions. When needed, the portable equipment from this lab can be set up on a cart and used in the plant to perform a variety of experiments, tests, demonstrations, and ionization checks.

If possible, the lab should be in a separate area from other lab facilities and should be equipped with a computer/terminal, static dissipative benches and chairs, storage cabinets, sink, exhaust hood, nitrogen, static grounds, and a dissipative floor. It should also be equipped with standard services such as electricity, water, vacuum, nitrogen and high-pressure dry air. The following is a list of the test equipment used in a General Laboratory.

- **Charge Plate Monitor** with printer
- **Charge Analyzer (Faraday Cup)**
- **ESD Simulators**
  - a. HBM
  - b. CDM
  - c. Zapper (HBM for systems)

- **Wide-Range Ohmmeter or Megger** with 2 NFPA probes
- **Buried Layer Tester**
- **Static Decay Time Tester** with controlled humidity chamber
- **Temperature/Humidity Meter**
- **Temperature/Humidity Recorder**
- **Air Ionizer**
- **Ion Current Meter and Residual Voltmeter**
- **SRM/RTG Meter** (may be separate meters)
- **Static (field) Voltmeter**
- **Personnel Voltage Tester**
- **Oscilloscope and/or curve tracer**
- **Power Supply** — modified with 10,000 megohm resistor in series with HV lead to charge people or materials. Adjustable from 0 to a minimum of 7500 volts.
- **ESD Monitor**
- **Grounded Cart**
- **Field Auditor's Kit**

Analytical/FMA Laboratory

The analytic laboratory, whether set up within a company or independently operated, should be equipped with or have available (through a related lab), the necessary facilities to do the following:

1. A Failure Mode Analysis of products
2. A chemical and physical analysis of materials
3. ESD properties of materials and devices.

Failure Mode Analysis of products: The Failure Mode Analysis laboratory should be equipped so that it can determine the causes of failures in integrated circuits or packaged devices. Following are some of the instrumentation needed to do this.

- Chemical etch systems/jet etcher
- X-ray radiographic imaging equipment
- Optical microscope/camera
- Scanning electrical microscope (SEM)
- Laser (to isolate circuit elements)
- Plasma etcher
- Micromanipulator/Curve Trace/Parametric Tester

Chemical and Physical Analysis of Materials: This laboratory should be equipped to do both organic and inorganic chemical and material analysis using normal methods. The following is a partial list of equipment that should be included.

- Auger microprobe
- Scanning electron microscope
- Secondary ion mass spectroscope
- Carbon/sulphur analysis equipment
- Gas mass spectrometer
- Gas chromatograph
- X-ray fluorescence equipment
- Atomic absorption spectroscope
- Optical emission spectrograph
- Laser emission microprobe

- Inductive coupled plasma emission spectroscope
- Organic carbon analysis equipment
- Fourier transform infrared spectrometer
- Ion chromatograph
- UV-visible spectrophotometer
- Thermal analysis equipment
- Wet chemistry equipment
- Metallographic examination equipment

ESD Properties of Materials: The ESD properties of materials should be analyzed using the appropriate material specifications issued by the following organizations:

- American Association of Textile Chemists and Colorists (AATCC)
- American National Standards Institute (ANSI)
- American Society for Testing and Materials (ASTM)
- Electronics Industries Association (EIA)
- Electrical Overstress/Electrostatic Discharge Association (EOS/ESD ASSN)
- International Electrotechnical Commission (IEC)
- National Electrical Manufacturers Association (NEMA)
- National Fire Protection Association (NFPA)
- United States Specifications (DOD-FED-MIL)

Equipment should include all of the equipment listed for use in the General Laboratory plus any special equipment needed under the listed specifications.

Instrument Descriptions

Instrument #1: Charge Plate Monitor (CPM)

This instrument was originally developed to monitor the balance of air ionizers and the rate of charge decay from a fixed voltage point (either positive or negative) to any desired ending point (usually 10 percent) of the original charge. These two tests give a measure of an ionizer's efficiency. Many other uses have been found for these monitors, thus making them very versatile piece of equipment.

To utilize the CPM for many of these extra uses, a connection to the monitor plate (usually through a wrist strap and cord without a resistor) should be made. Since the construction of instruments can vary with the manufacturer, it is impossible to cover all methods of making the connection. One method is used to drill a hole in the plate close to one edge and insert a banana jack. A wrist strap or other cord can then be connected to the banana jack. Connecting a wrist strap to the floating plate will change the capacitance of the instrument head and its final accuracy. However, for field measurements of floors, tables, and some demonstrations, complete accuracy is not necessary.

Using the Float Mode to Measure Triboelectric Charging: The CPM can be used to measure the triboelectric charging of floor finishes, rugs, mats, and ESD shoes or other footwear. Figure 6-1 shows how to measure the triboelectric charging of a floor finish. Using this method, select the floor to be tested, select the footwear to be used, place an insulating plate on the floor (for the individual to stand upon), and set up the monitor for a float test (time and scale will depend upon the type of unit used). Attach a wrist strap from the individual to the plate of the CPM, zero or ground the plate, and set it for the float mode. With the test running, have the individual step off of the insulating plate onto the floor and walk normally for the specified test time. During or after the test, depending on the recording method, the test information can be documented giving a record of the polarity and magnitude of the triboelectric charge generated by the test. This can be repeated for different footwear in order to better evaluate the charging of the floor.

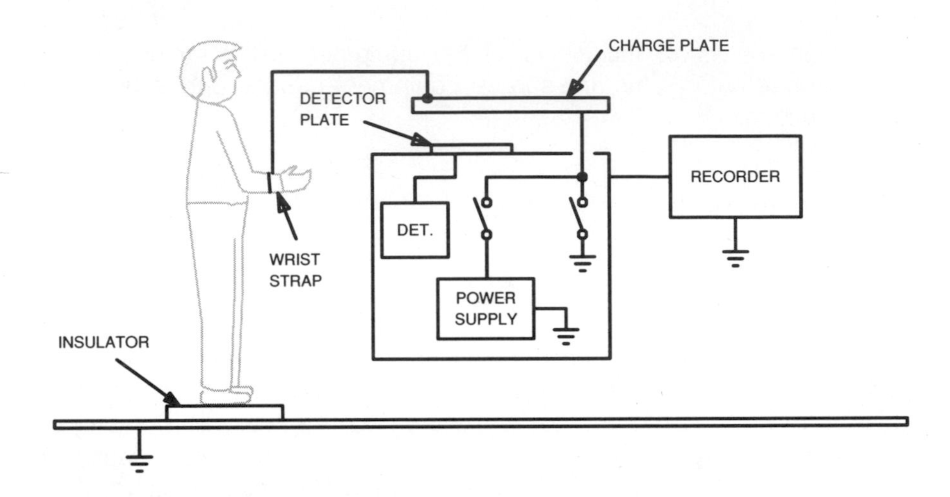

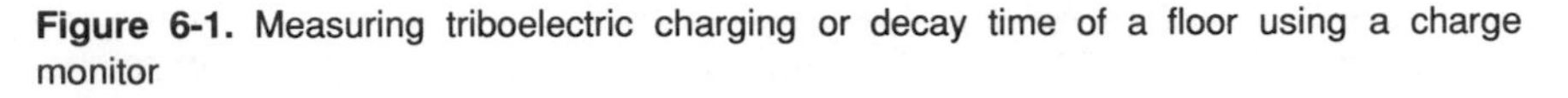

Figure 6-1. Measuring triboelectric charging or decay time of a floor using a charge monitor

Using the Float Mode to Demonstrate Induction Charging: Induction charging is a phenomenon that occurs when an ungrounded conductor enters a field and then is momentarily grounded. After the grounding (if the field is removed), the conductor will show a charge equal to the original charge impressed by the field but of the opposite polarity.

This can be shown by charging an insulator, then placing it on or near the plate of the CPM (which has been zeroed and placed in the float mode), and reading the potential. The plate is momentarily grounded, and the charged insulator removed. The plate will read an equal but opposite potential from that originally read. (See Chapter 11 for more details.)

Using the Decay Mode to Test: The CPM is also an effective instrument for measuring the decay time from charging an object to a given level and then grounding it (Figure 6-1). The instrument measures the decay times of items such as tabletops and mats, floor finishes and mats, footwear, bags or boxes through tables or floors, smocks, and other objects, or areas limited only by the individual's imagination.

For example, to test the decay time of ESD footwear on a given type of floor finish or tile floor, set the unit to decay mode, set the upper and lower voltage limits (that is, +5000 and 50 volts), and set a time limit (for instance, 5 seconds). Place an insulating plate on the floor, and have an individual who is wearing a wrist strap that is connected to the monitor plate of the CPM stand on it. Initiate the test and when the RUN indication appears on the CPM, have the individual step off of the insulator plate onto the floor. The unit will measure and, if desired, record the decay curve, elapsed time, and final voltage.

This method can also be used to check tabletops, mats, boxes, bags, and other materials for a decay time to ground or through a resistance (such as a mat or tabletop). See Chapter 10, "Purchasing Guidelines—Finding the Hidden Problems," for additional examples.

Instrument #2: Electrostatic Field Meter

An Electrostatic Field Meter is designed to detect and measure the voltage and field strength associated with static charges on objects and materials.

Evaluating an electrostatic charge is done indirectly by measuring the "electric field or potential" that a charge produces in its surroundings. Since the terms "field" and "potential" are related but different, an explanation may be in order.

- **Field**—The force exerted on a unit positive charge at the point where the reading is made. It is usually measured in kV/Meter with a non-chopped meter. (Primary user—Europe.)

- **Potential**—The measure of work done on or by a unit positive charge in moving it from a position of zero potential to the point where the measurement is made. It is measured in volts. A chopped meter measures volts directly. (Primary user—USA.)

Most of the older field meters, when activated, provide a continuous reading. While this is fine in the majority of cases, the presence of an ionizing field or a pulsating field can cause erroneous readings. To overcome this deficiency, some newer models are equipped with a chopper/stabilizer that continually resets a ground and allows accurate readings in ionizing fields.

Making accurate measurements with field meters depends upon maintaining the required distance from the object to be measured (this is specified by the meter and usually is dependent upon the voltage range) and having the object in a free space condition.

Free space means having the object to be measured in a stand–lone condition, free from any other object. For example, to measure a bag or box, it should be suspended in the air at least 1 foot from any tabletop or vertical surface and the tester should be sure that the object being tested is free from any charge field.

In addition, the meter itself should be grounded through a ground cord to equipment ground to avoid erroneous readings.

See Figure 6-2 for an example of how to measure the charge on a plastic bag. Suspend the bag in the air by having someone hold it with insulated tweezers, or suspend it in the air with a piece of tape. Position the grounded meter at right angles to the surface of the bag at the specified distance, and activate the meter. Move it around the surface to test for "hot" spots. Check the potential and record the data. Deactivate the meter and repeat with other bags as needed.

Instrument #3: Residual Voltmeter (RVM)

A residual voltmeter is an instrument that measures the residual or offset voltage (floating potential) of bench top ionizers. The unit works by measuring the excess ion current (nanoamperes) that is present on the sensor plate, which has been placed in an ion stream. The current is measured by passing it to ground through a very high impedance. The resulting voltage is displayed on a zero center meter. This measurement is primarily for bench type (convection or blower) ionizers and does not work well on a room system that depends on a laminar airflow.

For example, to measure the balance of a bench ionizer (Figure 6-3), place the grounded RVM in the air stream of the bench ionizer about 1 foot in front of the unit. Turn on the RVM and read the unbalance. One note of caution: Even if the unit shows no unbalance, it is necessary to check the output with an ion current meter. The reason? A unit that is not producing any ion current will show a zero balance.

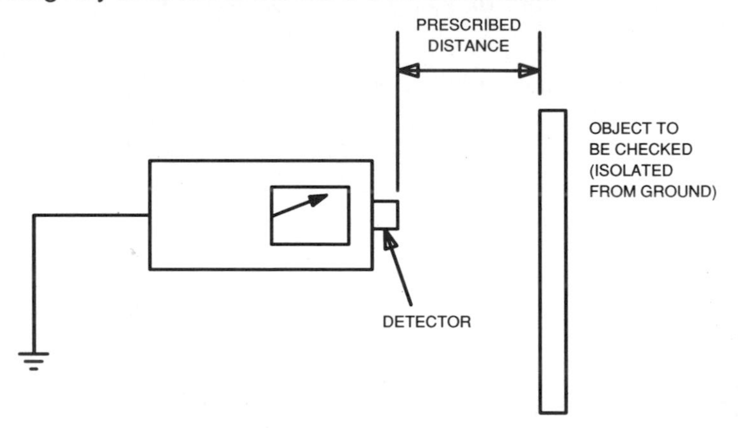

Figure 6-2. Measuring the charge on a bag using a field meter

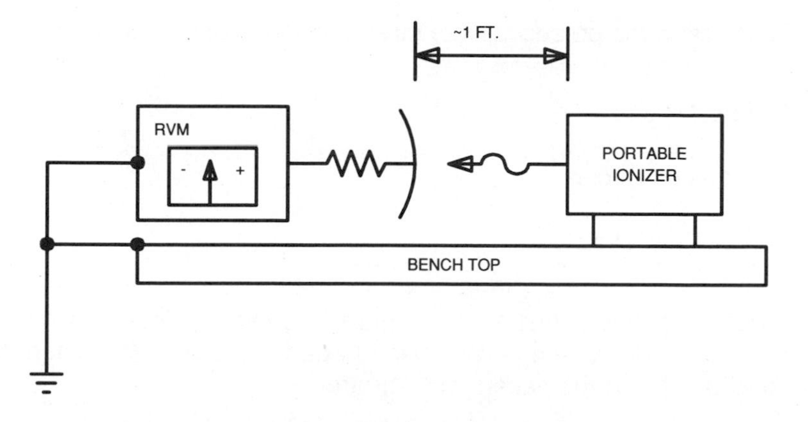

Figure 6-3. Measuring the balance of a bench ionizer using a residual voltmeter

Instrument #4: Surface Resistivity Meter (SRM)

Surface resistivity is defined as the ratio of DC voltage to the current that passes across the surface of a system where the surface consists of a square unit of area. In other words, surface resistance is the resistance between two opposite sides of a square and, as such, is independent of the size of the square or its dimensional units. The square can be formed by two electrodes of any length separated by a fixed distance equal to that length.

There are several types of meters that can be used to measure surface resistivity, ranging from laboratory-type high resistance meters through Megger meters to small portable units that give only decade-type readings. As there are a variety of meters, there are also several types and configurations of probes used. In general, two probes sit on the surface to be measured, within a small area, with current being passed between them (Figure 6-4).

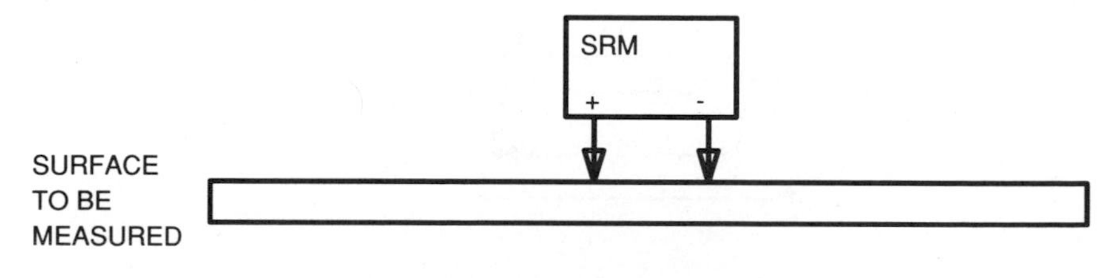

Figure 6-4. Measuring surface resistivity

Specifically, the probes fall into three configurations. These are:

- Parallel Probes
- Concentric Circles
- Three-Point Probes.

Parallel Probes: Illustrated in Figure 6-5, this configuration has two parallel electrodes of length **W**, spaced apart a distance **D**. When **W** is equal to **D**, it forms the pattern of a square.

$$\textit{Surface Resistivity} = \frac{\textit{Potential Gradient}}{\textit{Current per unit width}} = \frac{V/D}{I/W}$$

$$\textit{If } W = D, \textit{ then this becomes } \frac{V}{I} \ (\textit{Ohms/Square})$$

Concentric Circles: Illustrated in Figure 6-6, this configuration consists of a probe with two concentric circle electrodes, with diameters **D1** and **D2**, with the outer electrode being the ground and acting as a guard ring.

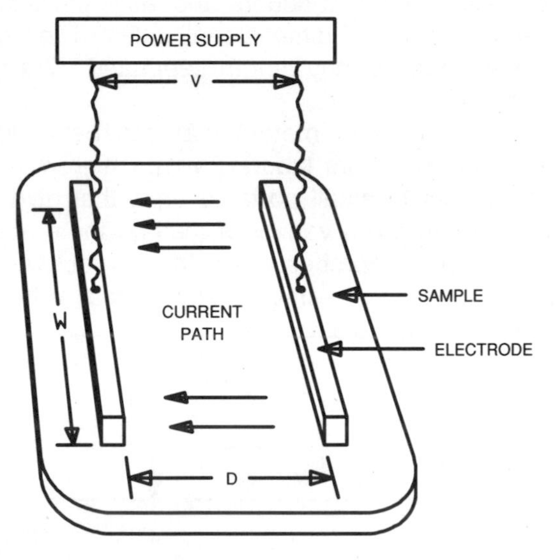

Figure 6-5. Parallel probe surface resistivity configuration

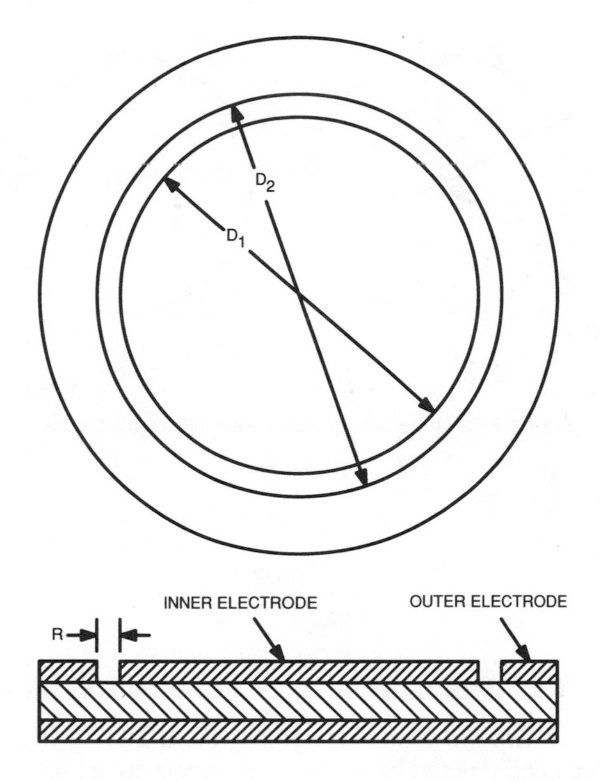

Figure 6-6. Concentric circle surface resistivity configuration

Surface resistivity is calculated as follows:

$$\sigma_s = \frac{\pi(D2 + D1)}{(D2 - D1)} \times R$$

where

D2 = Inner diameter of outer electrode
D1 = Outer diameter of inner electrode
R = Measured Resistance in Ohms

Three-Point Probe: This configuration, shown in Figure 6-7, has three powered probes at 120 degree angles on the perimeter of a circle, with a ground return probe at the center of the circle. The diameters of the probes and the distance from the inner probe to the outer probes are as shown in Figure 6-7. After calculating the number of squares from the three outer probes to the inner probe ($1/\sigma_s$ = APPARENT SQUARES), surface resistivity (σ_s) is calculated from the following

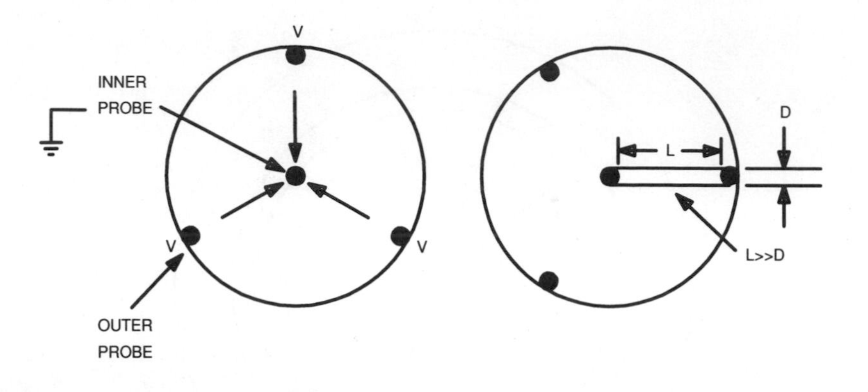

Figure 6-7. Three-point surface resistivity configuration

formula, where **L** >> **D**:

$$\sigma_s = \frac{6\pi R}{\log(16L^4/3D^{4)}}$$

Instrument #5: Resistance to Ground (RTG) Meter

A Resistance to Ground Meter is nothing more than a high-resistance ohmmeter that is connected between a grounded object and a ground so that the total series resistance can be read. (See Figure 6-8.)

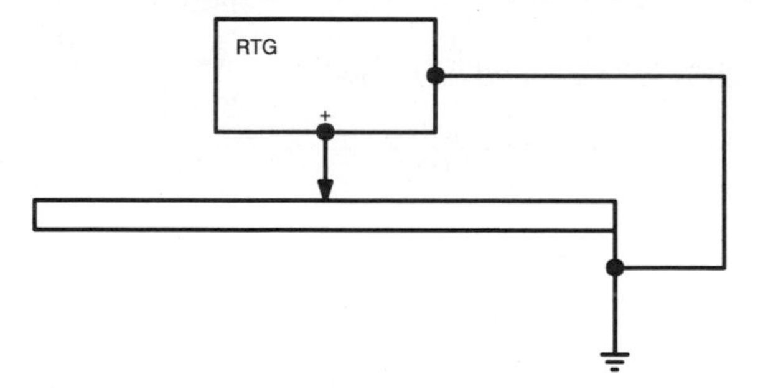

Figure 6-8. Measuring resistance to ground

Instrument #6: Combination SRM/RTG Meter

Several portable meters now combine the function of reading resistance to ground with a surface resistivity meter. This is accomplished within the meter by a switching arrangement that shifts the ground connection from the SRM probe to a plug. This plug is then connected to ground through a ground cable. These meters can have LED-decade readouts and/or an analog meter and usually read from 10^4 to 10^{12} ohms/square and/or ohms. The applied voltage can range from 30 to 100 volts. Caution should be taken to keep the contacts clean and dry. This type of meter is usually most accurate between 10^6 and 10^{10} ohms and, if compared to a point-to-point resistance, will read from 1-1/2 to 3 decades higher.

> ***Caution:*** The ground leads should be shielded to prevent parallel resistances where the ground line comes in contact with the surface.

> ***Note:*** When measuring resistance to ground, errors in readings may be observed if there is any RF or other interference riding on the ground line. Interference is more prevalent if there is a series resistance (1 megohm) in the ground line. In general these readings will be 1 to 2 decades higher than normal.

Instrument #7: Static Decay Time Tester With a Controlled Humidity Chamber

This instrument is used to test the decay time of a static charge from a given voltage to a predetermined percentage of that voltage under various humidity conditions. A sample of the material to be tested is placed in a holder and charged, and then a ground is applied. The decay is measured by a noncontacting voltmeter with the time displayed on an indicator. (See Figure 6-9.)

Samples of materials, boxes, gloves, tapes, and almost anything else that can be cut into a 3"x5" sample can be tested.

The humidity chamber is used to precondition the samples to meet the various standards and then to maintain them at the proper humidity during the test.

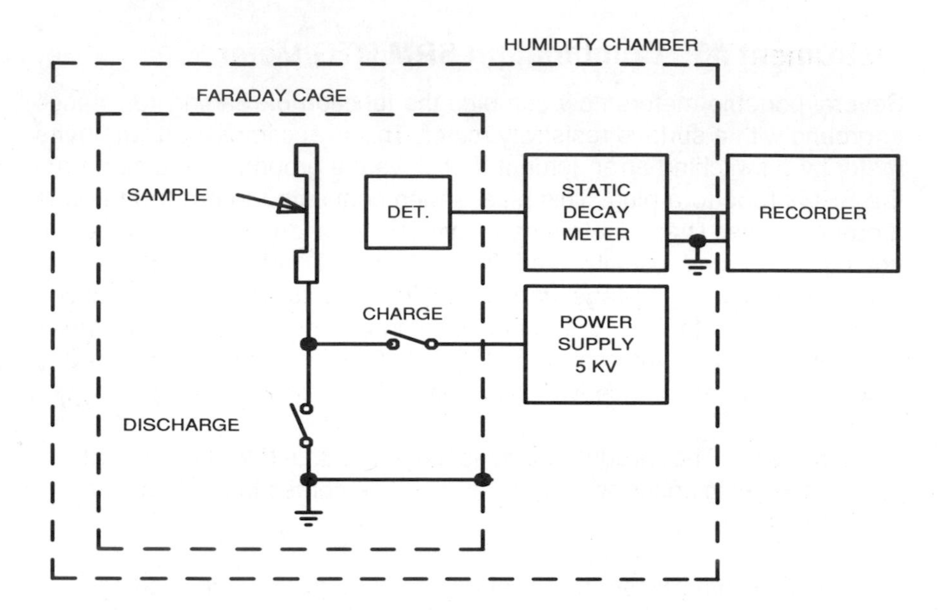

Figure 6-9. Measuring the static decay time of material samples

Instrument #8: Megger Meter (High Resistance Ohmmeter) with NFPA Probes

This is a high-resistance ohmmeter that applies a voltage, usually 10 or 100 volts, and is used with the NFPA-style probes to measure point-to-point resistance or resistance to ground of such items as tabletops, mats, and floors or floor finishes. This is different from a surface resistivity reading in that the measurements are made in ohms and not in ohms/square. Correlation studies between the two types of measurements have shown that surface resistivity can run 1-1/2 to 3 decades higher than the point-to-point resistance s.

Point-to-point resistance is becoming more meaningful for material testing since the change in ASTM D257, which restricts that standard to resistivities in the insulative range ($>10^{12}$), and because EOS/ESD Draft Standard #4 has adopted point-to-point test methods.

Point-to-point resistances on tabletops can be measured by using the procedures outlined in EOS/ESD Standard #4. Place the probes 10 inches apart, connect to the meter, and measure the resistance. (See Figure 6-10.)

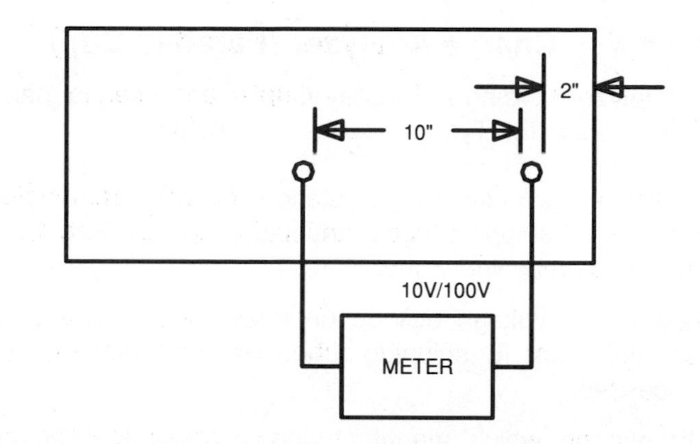

Figure 6-10. Measuring the point-to-point resistance of tabletops

Resistance to ground of tabletops can also be measured by using the procedures outlined in EOS/ESD Standard #4. Place one probe on the tabletop and ground the other side of the meter. (See Figure 6-11.)

The point-to-point resistance of a floor or floor finish can be measured by placing the cleaned probes 30 inches apart on a portion of the floor that has been cleaned of dust and dirt. Connect the probes to the meter and apply 100 volts. Read the point-to-point resistance (per NFPA-99-89). (See Figure 6-12.)

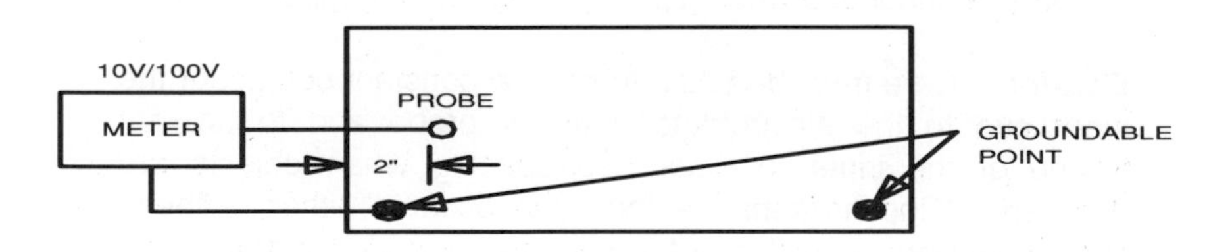

Figure 6-11. Measuring the resistance to ground of tabletops using an NFPA probe

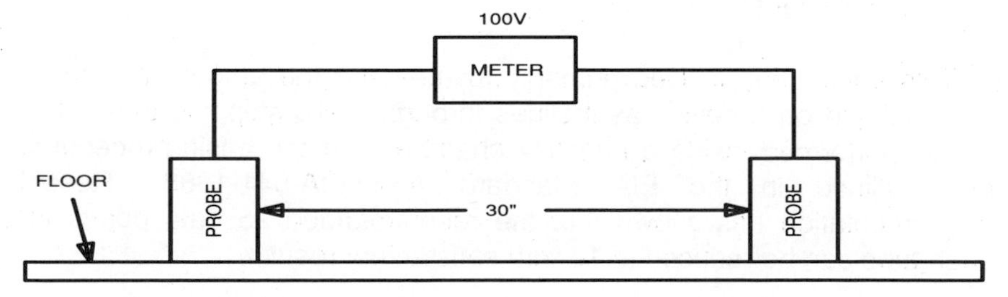

Figure 6-12. Measuring the point-to-point resistance of a floor

Instrument #9: Charge Analyzer (Faraday Cup)

A Charge Analyzer utilizing a Faraday Cup is an essential part of static identification and control. Typical applications include:

1. Testing the effectiveness of static prevention materials such as antistatic IC shipping tubes, antistatic bags, packaging materials, automated machinery, etc.
2. Testing the voltage buildup on integrated circuits as they are removed from IC shipping tubes or automated manufacturing processes.
3. Discovering which manufacturing process is producing ESD damage to components.
4. Measuring the total charge on a PWB assembly.

The Faraday Cup works on the principles set up by Michael Faraday. He discovered from experimentation that charges placed on an insulated hollow conductive object appear to reside entirely on its outer surface. Using this finding, he constructed a (Faraday) cup (with one cup located inside another) as shown in Figure 6-13. The internal cup is connected to a capacitor in an electrometer and the external cup is connected to ground, thus providing a Faraday shield for the inner cup. An electrometer is then used to measure the amount of charge that is shared with the inner cup when a charged object is dropped into it.

> ***Caution:*** Care must be taken to ensure constant contact of the inner cup to the electrometer lead or probe and to prevent flexing of the inner to outer cup spacing when objects are dropped in. Inconsistent readings can occur if either of these should happen. A quick test for this can be made by dropping a small, uncharged, steel ball into the cup and making a series of measurements.

This equipment is used primarily to measure the amount of charge accumulated on a device as it slides through an IC shipping tube. The test is performed inside a humidity chamber and the basic procedures are outlined in the EIA Standard ANSI/EIA-541-1988. Recent experimentation has shown that the recommended 25 data points for each tube can be reduced to 10 with satisfactory results.

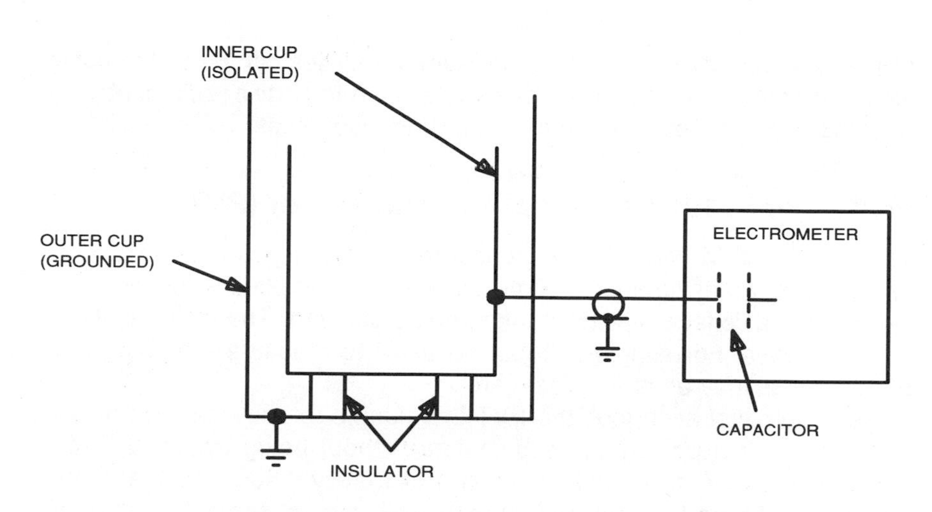

Figure 6-13. Cross section of a Faraday Cup

It is important to note that many variations may occur during testing that cannot be documented in any test specification. Among these possibilities are:

1. Variations from test employee to test employee. In all cases, the employee must be grounded.
2. Variations in the positioning of the IC shipping tubes over the Faraday Cup.
3. Variations from clothing. The Faraday Cup can pick up the field generated by clothing. It is recommended that a static shielding coat be worn.
4. Variations from the Humidity Control Chamber. If the chamber is equipped with solid gloves, the gloves can cause variations. One solution is to replace the gloves with iris ports and use bare hands.

Although it is virtually impossible to eliminate all of the variables in this test, it is helpful to identify them. Some variations can be minimized by consistent treatment.

Another use of the Faraday Cup is to test automated manufacturing facilities such as conveyors for the charging of PWB assemblies. The PWB assemblies should be dumped directly from the conveyor into the cup or placed in the cup using some type of insulated pliers in order to prevent the boards from discharging. Other facilities can be tested using similar methods. Tests of this type can be used to certify equipment prior to using it to minimize ESD damage on production lines.

Instrument #10: Personnel Voltage Tester (PVT)

The Personnel Voltage Tester measures the static potential on people using a direct measurement voltmeter. When a test plate is touched and depressed, a voltage reading is taken and displayed. The tester is useful in many ways. For example, it can be used to measure the charge on personnel as they enter any given area.

As a demonstration tool, the PVT is valuable in showing the amount of charge that a person can accumulate without being aware of it (the human sensation threshold being approximately 3500 volts) and the protection offered by wrist and shoe straps. As an example, a person entering an area can measure his charge, then, after putting on a wrist strap, remeasure the charge. It should read approximately zero volts. The same type of effects can be measured with the other ESD control items for personnel.

Instrument #11: Air Ionizers

An air ionizer is a device used to neutralize charges on insulators or other objects that are incapable of being discharged through grounding. The ions are generated by either a weak alpha–\mitting nuclear isotope or from emitter points using a high voltage source. The high voltage units come in several source forms, from alternating current to direct current and several types of pulsed direct current. Ionizers come in many sizes ranging from small bars to portable bench top units to full room installations. All high-voltage units require some form of periodic maintenance to clean the emitter points and to check output, balance, and discharge time. No one type of ionizer can do all jobs, and (for safety reasons) high-voltage units should never be used in an explosive atmosphere.

Instrument #12: ESD Simulators

ESD simulators are used to replicate the effects of the ESD phenomena that threaten microelectronic devices during manufacturing and assembly. The simulators are based on the HBM and the CDM as discussed in Chapter 5.

Human Body Model (HBM) Simulator for Devices: The HBM simulator is the best-known simulator used in the industry today. It produces an electrical pulse that resembles that of a charged person touching the lead of a grounded device. The current industry standard for the RC network is a 100 pF capacitor and a 1500 ohm resistor.

There are two different types of commercial HBM ESD simulators: manual and automatic. Both types basically employ the RC network mentioned above and a high voltage power supply.

The automated simulators provide a fully computer-controlled pin selection and voltage setting. The manual set requires that the employee select the pin to be tested, adjust the power supply to the desired voltage level, and initialize the test for each zap. Both types of simulators require a specific test board to accommodate each device package layout. In addition, the automatic sets usually come with a built-in curve tracer to allow device characterization, while most manual sets only supply a connection port for an external curve tracer.

Charged Device Model (CDM) Simulator: At the present time, no commercial CDM simulator is available. This is due to the lack of an industry standard test method. However, several manufacturers have developed CDM testers for their own use. The most commonly used "homemade" CDM tester is the one introduced by Bossard et al (1980).[29] An IC lying upside down on 1/32-inch thick insulating material is charged by applying voltage through a high resistance probe and then discharge to ground through a relay. The use of the relay severely limits this approach. (See Figure 5-7.)

A more efficient and flexible simulator, which takes advantage of the principle of static induction (Chapter 3) has been described by Renninger et al.[52] This field-induced charged-device model (FCDM) simulator raises the DUT to a high voltage using the field of a noncontacting electrode. This process helps to avoid premature device stressing and permits a higher throughput. In addition, the tester discharge the charged device by direct contact with an electrode in air rather than through a relay, which more closely replicates the real situation. (See Figure 5-8.)

HBM ESD Guns for Circuit Assemblies and Systems: Electronic subassemblies and systems are usually tested using an ESD "gun." These guns are used to simulate a charged person or object touching operating equipment where the discharging pulse may upset the system and trigger an unexpected system failure. The gun can be hand-held or tripod-mounted. It comes with a control power supply unit and interchangeable discharging networks and various profile tips for different applications. The control unit can be programmed to perform in both repetitive and single-shot modes. The user decides which discharge network to use, sets the power supply to the desired voltage level, and pulls the trigger to initiate the test. Considerable controversy exists over whether the discharge should occur in air (more realistic) or after contact (more repeatable).

Points To Remember

- Testing must be an integral part of any ESD program.
- Testing must be tailored to meet the needs of the company or organization.
- Testing must be done in accordance with the appropriate standards in order to be viable.
- Care must be taken to properly ground and position the test equipment.
- The effects of charge suppression must be considered when using field meters.
- Consideration must be given to providing the most cost-effective equipment for the job to be done.
- Equipment must be easy to use, maintain, and calibrate.
- For demonstration purposes, analog meters are preferred.

Chapter 7

Realistic Requirements

The ESD control techniques and requirements in this chapter are based on research carried out at AT&T Bell Laboratories, detailed manufacturing studies such as those in Chapter 4, and the author's eleven years of experience in manufacturing environments. These requirements were first published in a handbook at the AT&T manufacturing facility in North Andover, Massachusetts. Over time, they have been thoroughly reworked by the AT&T Corporate ESD Committee so that they now represent a consensus held by all AT&T locations. In fact, they are now the accepted requirements at all AT&T manufacturing facilities and appear in their final form in the AT&T *Electrostatic Discharge Control Handbook*. They have been carefully formulated to define minimum standards that will still allow maximum flexibility at each location. As such, they have been successfully used at locations having as few as 50 employees and at others with as many as 11,000. That is to say, they have been proven realistic and suitable for large or small operations.

The AT&T handbook requirements that follow provide a common answer and a realistic solution to ESD control for the electronics industry. They can be used as is, with confidence, or modified to accommodate specific needs. One hundred percent compliance with these handling procedures is the ultimate goal of the ESD control program, and compliance can be measured with the auditing techniques described in Chapters 8 and 9.

These AT&T handbook requirements were prepared with training in mind and with the understanding that only realistic requirements can be enforced. When engineering takes technology, training needs, and human nature into account, the result is a program of realistic handling requirements that are manageable as well as effective.

There is a tendency when developing a control program to over-engineer systems and to over-supervise employees. The result is handling systems and protective devices that are expensive and unmanageable. The handling requirement generated becomes either too rigorous or too difficult for bench personnel to execute faithfully and regularly. As a result, employees grow frustrated. In the end, their resistance to these handling requirements dooms the entire program to virtually certain failure. When, on the other hand, requirements are based on a thorough understanding of ESD technology, extensive research, and practical experience, there is no need to resort to unnecessary precautions.

There is no doubt that some overprotection and redundancy is necessary and cost-effective in the control of damage from ESD. Redundancy provides added layers of protection to ensure that sufficient safeguards are in place and provides some insurance against human error. The problem for an ESD committee is to determine how much redundancy is needed.

Sensitivity Classification

One reason that our requirements have proven to be cost-effective in such a wide variety of applications is that ***the requirements vary with device sensitivity***. Older, less sensitive applications require minimum precautions at minimum cost. On the other hand, the requirements for ultrasensitive devices specify more extensive precautions. These more demanding requirements are still realistic, providing only as much insurance as is absolutely necessary.

The requirements are based on an area sensitivity classification system which lists five classes of sensitivity:

AREA SENSITIVITY CLASSIFICATION

- Class 0 areas contain devices with ESD thresholds ranging from 0 to 199 volts
- Class I areas range from 200 volts to 499 volts

- Class II areas range from 500 volts to 1999 volts
- Class III areas range from 2000 volts and up
- Class IV areas do not contain devices that are sensitive to ESD damage

Based on these classifications, each area in a manufacturing facility is classified according to the most sensitive device handled in that area. Consequently, PWB assemblies are classified according to their most sensitive component. For example, a device with an ESD threshold of 100 volts would be assembled or produced in a Class 0 area, and all other devices in that area would be handled by the same Class 0 requirements. A device with a minimum threshold of 10,000 volts may be produced in a Class III environment. Operations that involve hardware only, in which there are no ESD–sensitive components, would take place in a Class IV area. Examples of this classification include backplane wiring or faceplate assembly.

This system of area classification actually simplifies training and communications within a manufacturing plant. It allows for greater protection to be engineered into the handling of more sensitive devices without requiring employees to learn more complicated handling requirements. The added protection is provided by engineering, with the facilities that they specify based on the sensitivity of the devices. This makes it possible to train employees with techniques that are common to all areas. For instance, Class III does not require a dissipative work surface whereas Class II does. The devices in Class II are better protected, and yet employees in both areas use the work surface in the same manner. It is the work surface that provides the added protection and not a change in procedures.

With few exceptions, the system is designed to offer better protection for more sensitive devices and to minimize the impact on training. An employee can enter any shop regardless of the sensitivity classification, use the facilities present in that area, and know that he/she is complying with proper procedures. ***Consequently, every employee can be taught that the same techniques apply in virtually all cases, regardless of the sensitivity classification.***. Yet even ultrasensitive devices end up with a good deal of added protection engineered into the facilities in which they are handled.

The control techniques applied to each classified area are listed in Table 7-1 and each one is required. All of them are defined in the pages that follow, and some have additional explanatory comments. Techniques are designated as either "R," always required, or "S," required when specified by engineering. The latter is an option exercised by engineering when unusual hazards exist that may require more than the minimum precautions or when items such as tote trays wear out.

A review of the table indicates that the Class III precautions are basic: grounding employees, training, certifying the facility, properly transporting the product in carts, and packaging. These are minimum precautions. Typically, this less-stringent set of requirements is applied to older robust technologies that have been in place for a considerable length of time. Class II areas require some additional precautions. Class I areas require even more stringent regulations, and Class 0 areas generally demand the most precautions.

For instance, when a tote box is used in a Class I shop, it is required to be static dissipative, whereas in a Class II shop it is not. This is the primary difference indicated in Table 7-1 between Class I and Class II areas. However, even in Classes II and III, engineering may specify that tote trays be replaced by static dissipative material, generally on a wear-out basis. In this way, a realistic, carefully controlled continuum of controls that have little impact on training is provided.

Control Techniques

Three Basic Rules for ESD Control

Three basic rules apply to all situations regardless of the area classification:

1. Assume that all electronic (solid-state) components and assemblies are sensitive to ESD damage.
2. Never touch a sensitive component or assembly unless it is properly grounded.
3. Never transport, store, or handle sensitive components or assemblies except in a static-safe environment.

A static-safe environment for electronic components involves:

- Controlling static charge generation.

Table 7-1. Minimum-Requirement ESD Control Techniques Specified by Class

Item	Area Classification (Note)				
	0	I	II	III	IV
Control Techniques	Device Sensitivity				
	0-199 V	200-499 V	500-1999 V	2000 V+	None
Personnel Awareness/Training	R	R	R	R	
Auditing Compliance To ESD	R	R	R	R	
Personnel/Facility Certification	R	R	R	R	
Personnel Grounding	R	R	R	R	
Transporting Products:	R	R	R	R	
Carts	R	R*	R	R	
Static-Safe Packaging	R	R	R	R	
Static-Safe Bags	R	R	R	S	
Static-Safe IC Shipping Tubes	S†	R	R	S	
Static-Safe Tote Boxes	R	R	S	S	
Dissipative Table Mats, Tabletops	R	R	R	S	
Dissipative Floor/Finish	R	S	S	S	
Edge Connector Shunts	R	R	R	S	
Extraordinary Measures	R	S	S	S	
Other Controls (As Required):					
Air Ionizers	S	S	S	S	
Antistatic Smocks	S	S	S	S	
Conductive Packaging	S	S	S	S	
Conductive Foam	S	S	S	S	

Notes: R = Always required when used.
S = May be specified by Engineering to provide additional protection for unusual hazards.

* See the subsection on Carts for added Class 0 and I requirements.

† See the subsection on Static-Safe Bags, Shipping Tubes, and Tote Boxes.

- Eliminating charges wherever they exist. How these charges are dealt with depends on whether the charged object is a conductor or a nonconductor.

- Having a common grounding philosophy. This is essential where all conductors in the workplace are grounded, including people, table mats, floor mats, tote trays, wrist straps, heelstraps, etc. People must always use grounded wrist straps or conductive footwear in conjunction with grounded conductive flooring.

These steps will ensure that static potentials within the system remain near zero. Handling and grounding techniques combined with an awareness program that enlightens personnel about the hazards of ESD are the most effective methods for eliminating ESD damage.

Definitions

Static-Safe

Conditions or materials that provide adequate protection for the product sensitivity in the intended application are considered to be static-safe. However, the ideal material often does not exist, and it is virtually impossible to prevent some degree of charging. Therefore, trade-offs are generally involved. Engineering judgment is necessary to consider these trade-offs and to determine whether a material or condition is static-safe.

Static Conductors and Nonconductors

Conductors, such as metal or carbon-impregnated materials, allow electrons to flow and can eliminate a charge when connected to ground. Nonconductors (expanded polystyrene [EPS] and other plastics) will not allow a charge to flow. Therefore, ***grounding an insulator is useless.*** Eliminating nonconductors from the workplace is the best solution, but if this is not possible, always avoid direct contact between the sensitive products and the nonconductors. The next best solution is using ionized air to neutralize the surface charge (see Chapter 12 for additional discussion).

Surface Resistivity

For steady current flowing along a surface, surface resistivity is defined as the ratio of the electric field (v/m) to the surface current density (amps/m). The unit of surface resistivity is ohms per square. For a uniform current density, surface resistivity equals the ratio of the voltage drop across a square region of any size to the total current flowing through that region (also see pages 49, 50 and 105. Two categories of surface resistivity characterize most ESD-combative materials:

- Conductive (0 to $<10^5$ ohms per square)
- Static-dissipative (10^5 to $<10^{12}$ ohms per square).

Conductive Materials

Conductive materials may or may not be antistatic and are the quickest to dissipate charge. They are used to shield some highly sensitive products. However, they do offer an element of considerable risk of ESD damage based on the charged-device model. Therefore, one should always avoid contact between the conductor of a static-sensitive product and conductive surfaces. If this contact is a process requirement, use every precaution to ensure that the product is not charged before contacting a conductive surface and that the conductive surface is grounded. When a charged product touches a conductive surface, the rapid discharge may result in ESD damage.

Conductive packaging materials for shipping are generally not used because of the higher costs, and very few products require shielding for either handling or shipping (see Chapter 12 for additional discussion).

Antistatic Property

This property refers to the prevention of triboelectric charge generation. It will effectively minimize the production of a static charge when materials are separated from another surface. This property is not a dependent function of material resistivity or of static decay performance. In other words, a material could be conductive or static-dissipative and not antistatic (see Chapter 12 for additional discussion).

Static-Dissipative Materials

In general, static-dissipative materials that are also antistatic are preferred to conductive materials because charge dissipation occurs at a safe rate—neither too fast nor too slow. Both categories have appropriate applications when properly used. For example, tote boxes could be lined with dissipative material on the inside where they contact the product but be conductive on the outside for rapid charge dissipation or shielding (see Chapter 12 for additional discussion).

Static-Generating Materials

When ESD protection is required, all static-generating materials, such as untreated plastics, should be removed from the workplace. This is not always possible since the materials may be an important part of an operation. For example, it would be virtually impossible to eliminate such items as static-prone clothing, calculators, terminals, and faceplates. To minimize ESD damage, follow these requirements:

- All nonessential plastics, such as coffee cups, food wrappers, and pocketbooks, must be removed from the top surface of the workstation when ESD-sensitive products are present. Additionally, direct contact with these materials must always be avoided.

- Essential plastics (clothing, calculators, CRT terminals, faceplates and part bins) may remain at the workplace (preferably toward the rear of the workstation). Direct contact between these plastics and ESD-sensitive products must be avoided unless it is a process requirement (such as the assembly of plastic faceplates to PWB assemblies). The ESD-sensitive parts may be stored in static-prone part bins and containers only if they are ESD protected. The protection may consist of separation by means of conductive foam, static dissipative bags and foams, conductive inserts, or topical antistatic sprays. However, highly sensitive (Class 0 and Class I) components require additional protection. In Class 0 and Class I areas, modifications to part bins or holding fixtures must also include a continuous path to ground if bare parts are to be loaded or unloaded.

- All new purchases should specify ESD-safe materials whenever available.

This definition also establishes the requirements for dealing with static–generating materials and impacts on virtually every operation. These requirements are realistic solutions that were developed over several months with considerable input from employees, supervisors, engineers, the ESD committee, and Bell Laboratories. They have proven to be not only realistic, but also highly effective in preventing ESD damage.

What makes the requirement realistic is the recognition that it is virtually impossible to completely eliminate static generators. Therefore, it is necessary to acknowledge their presence and to manage them accordingly. For instance, it is reasonable to expect that all unnecessary plastics, such as shopping bags, be removed from the workplace. On the other hand, it is impossible for the employee to eliminate plastic faceplates on PWB assemblies. When presented in this fashion, the employee can readily accept the difference. Therefore, it is realistic to expect strict compliance. The result is a workstation that is free of any unnecessary risks.

Regarding the necessary risks of static generators, engineering judgment is needed to determine whether the no-direct-contact requirement is sufficient. In applications where strong fields are present, an ionizer may be necessary. However, in most instances, the requirement is sufficient as written.

ESD-Insensitive Devices

Some solid-state components are, in a practical sense, immune to ESD and may be classified as ESD-insensitive (Class IV). However, given sufficient abuse, any device will fail. Additionally, design and process changes can result in sensitivity changes. ***Therefore, classifying a solid-state device as ESD-insensitive is not recommended and should be done only with extreme caution.***

> ***Note:*** Permitting some solid-state components to be considered ESD-insensitive was the result of a compromise reached during the development of the AT&T *Electrostatic Discharge Control Handbook* and is a true reflection of a consensus standard. The purists were inclined not to permit this definition, but after considerable debate, it was found to be the only realistic solution. Some of the more robust product lines were experiencing very high yields and outstanding field return performance. In these instances, it would be very difficult to justify the expense of ESD controls. Thus it was agreed to

include this definition and permit exceptions. However, these exceptions are very rare and are not recommended unless they can be justified with supporting data.

Minimum Requirements

The following control techniques are minimum requirements and are included in the AT&T ESD Control Handbook. They are repeated here for convenience. For example, the control technique for an ***ESD training program*** is stated just as it appears in the handbook. Each one of the techniques is listed in Table 7-1, where the application is defined on the basis of the sensitivity classification. When appropriate, a brief discussion giving further insight or clarification will follow the control technique.

ESD Awareness Training

The success of an ESD control program depends on all personnel fully understanding the extent of the ESD problems. An ESD awareness training program is required for engineers, shop employees, and management. It is the responsibility of the engineering or training organization to develop and implement the ESD awareness program.

ESD Training Program

In-depth training is required and is a critical factor in the success of an ESD control program. The training procedures can be developed by the engineering or training organization to suit specific needs.

Discussion: Special emphasis should be placed on training engineers, maintenance technicians, and first-line supervisors. They must be trained in the proper ESD control procedures and must set an example by following ***all*** of the procedures ***all*** of the time, even when handling a product that is to be scrapped. See Chapter 11 for an in-depth discussion of training.

Auditing Compliance to ESD Procedures

A program for auditing compliance with ESD procedures shall be established for all ESD-sensitive areas. This program could be based on the AT&T *ESD Inspector's Guide* or an equivalent inspection method for verifying compliance with this handbook.

Discussion: This is another example of a realistic requirement that is suitable for an entire company, large or small. It clearly states that each location must have an auditing program in place and yet the specific techniques that are permitted vary from one location to another. This flexibility is essential especially in a large company. On the other hand, the corporate ESD Committee has gone one step further and developed an auditing technique that has been unanimously adopted. It has worked successfully at a number of locations and is described in detail in Chapters 8 and 9.

Checks and Maintenance

Control products, such as wrist straps, heelstraps/toestraps, tote boxes, dissipative mats, and tabletops, must be periodically checked for conductivity and protective isolation. The procedures used and the interval required are to be established by the responsible engineer. This decision should be based on manufacturer recommendations and local experience. However, it is suggested that wrist straps be tested daily and that mats and tabletops be tested weekly.

Discussion: Maintenance is a vitally important aspect of an ESD program and yet is often taken lightly. If a company is unwilling to pay attention to such details, it should save its money and not get involved in an ESD control program at all. Equipment readily becomes ineffective without maintenance, the whole program becomes, as a result, nothing more than an added expense.

Personnel Certification

All employees who have contact with ESD-sensitive components and assemblies should undergo an ESD Certification Training Program that teaches proper handling techniques. They should be tested by a valid testing method that will verify the accuracy of their ESD knowledge.

Note: Refer to Chapter 11 on training for a detailed discussion of this certification training.

Facility Certification

Automation makes the ESD qualification of handling, testing, and manufacturing equipment increasingly important. Automated facilities must be designed with two purposes in mind. One is to prevent damage to the items being manufactured or handled. The other is to protect the facility itself from ESD damage or malfunction.

New facilities should meet the following criteria listed in the following subsection, and the design of older facilities should be reviewed in accordance with the same criteria.

ESD Acceptance Criteria for Equipment Associated with Manufacturing, Handling, and Testing

For final acceptance, equipment should be qualified for ESD-safe operation in accordance with the following requirements:

- All conductive materials must be grounded.
- Movement must not cause excessive triboelectric charging of PWB assemblies or components. Refer to EIA (Electronic Industry Association) Standard No. 541 for triboelectric test methods.
- All surfaces in contact with PWB assemblies or components must be static-dissipative or conductive per EIA No. 541.
- The manufacturing facility shall not be subject to ESD damage or malfunction due to ESD. Refer to IEC (International Electronic Commission) Publication 801-2 for testing methods and requirements.
- The facility shall be equipped with provisions for convenient wrist-strap grounding of appropriate personnel. The ground point shall be identified.
- Work surfaces shall be covered with a grounded approved static-dissipative mat or tabletop laminate.

These criteria are complex, both technically and administratively, and are discussed in depth in Chapter 13, "Automation."

Personnel Grounding

Personnel are required to be properly grounded whenever they will be touching a sensitive component or assembly. In addition, operating equipment may be sensitive to ESD. In these instances, personnel grounding is required during such manufacturing steps as system testing. Grounding is not required when sensitive items are in approved transporting containers (tote boxes, IC shipping tubes, factory packaging, etc.).

The method used for grounding personnel must include either a wrist strap or conductive footwear. When seated, each involved employee must wear a wrist strap connected to a grounding system. When standing, conductive footwear may be used with conductive or dissipative flooring and floor finishes as an alternative. The grounding system (typically equipment ground) must be common to all ESD control facilities (tabletops, floor mats and wrist straps).

> ***Discussion:*** Grounding of personnel is a baseline defense for all ESD activities. It will solve nearly 80 percent of the problems in a cost-effective manner and produce yield improvements even on less sensitive product lines. However, it is necessary to anticipate the possibility of human error or equipment failure. Layers of protection establish the insurance necessary to minimize this possibility. Items such as static dissipative table mats, tote trays, and bags become an integral part of the system and must be designed accordingly. In this way, if one precaution fails, one of the added layers will minimize the possible losses.
>
> Furthermore, grounding through footwear is only an option when one is standing. In other words, if seated, there is no guarantee that either foot will be on the ground; therefore, wrist straps are essential and required once a person sits down. There is one drawback; people who stand most of the time tend to forget to use the wrist strap when they sit down. This is one of the human factors that must be anticipated in both the training and enforcement aspects of the program.

Wrist Straps

The sole purpose of wrist straps is to ground personnel. Typically, the straps are flexible wrist bands with minimum of a 1-megohm resistor in the ground cords, and they must contact the skin to be effective. Wrist straps with alligator clips may be used to connect to ground in areas not equipped with banana plug connectors. Continuity from the point of connection to ground must be maintained at all times. Wrist straps should never be alligator clipped to the edge of any table mat. The wrist strap must be connected directly to equipment ground.

Footwear

Several types of conductive footwear are available, all of which must be used with conductive or dissipative flooring. In all cases, footwear must be worn on ***both feet***.

> ***Discussion:*** Conductive footwear is effective only when used with grounded flooring such as conductive or dissipative floor mats, floor tiles, or floor finishes. Some companies specify footwear, such as heelstraps, on only one foot as a cost cutting measure. In reality there is no guarantee that the correct foot will be the one on the grounded flooring. Therefore, it is possible to be ungrounded while working on sensitive items. Furthermore, it is unrealistic to expect proper training as the answer to this issue. Requiring conductive footwear on both feet is the only viable solution.

Heelstraps/Toestraps

These devices are used to ground personnel at standing workstations. The typical heelstraps/toestraps are conductive straps that can be used with most types of shoes. Heelstraps/toestraps may be used with steel floors only if a minimum of 1 megohm of resistance is provided between the employee and ground. In all cases, heelstraps/toestraps must be worn on ***both*** feet.

Bootstraps

Bootstraps are alternatives to heelstraps. Bootstraps must be worn on footwear that extends too high to accommodate the standard heelstrap. They may also be used on conductive (steel) floors if the bootstraps are equipped with a minimum of a 1-megohm resistor. Bootstraps are not disposable and are much heavier than heelstraps. Employees should consider wearing footwear that is compatible with heelstrap design.

Conductive Boots

Conductive boots are disposable, fit over street shoes, and contain a conductive strip to ground the wearer. They are not recommended for use with steel floors.

Conductive Shoes

Typical conductive shoes have conductive undersurfaces that electrically connect the user's feet to the walking surface through a minimum of 1 megohm of resistance. They offer a more permanent and reliable alternative to heelstraps and bootstraps.

Conductive and Dissipative Floor Mats

Conductive and dissipative floor mats are used primarily to ground personnel. Like table mats, floor mats must be grounded. Refer to EOS/ESD Draft Standard #6.0 - "Grounding" for methods and safety considerations. It is recommended that footwear be specified that will provide adequate resistance to protect personnel from shock hazard.

> ***Discussion:*** Primarily, floor mats are used to provide personnel grounding when standing is required for prolonged periods of time in a confined area. Therefore, antifatigue mats are desirable. They can also solve those safety problems that a wrist strap cord may present when used in the proximity of moving machinery. If mobility is required over a wide area, conductive flooring or a conductive floor finish may be a better alternative.
>
> Floor mats are not typically needed at seated work positions because wrist straps are required when seated and are sufficient protection for most applications. The one possible exception to this would be for Class 0 applications where such redundant protection may be an appropriate safeguard.

Stationary Work Positions (Benches, Test Sets, Etc.)

At seated work positions, both the work surface and the employee must be connected to the common grounding system. The employee must wear a grounded wrist strap to be properly grounded.

Holding fixtures at the workstation should be made of static-safe materials and should directly contact the dissipative surface (especially in Class 0 or Class I areas). Where possible, all nonconductive materials must be eliminated from the work area. Where it is not possible to restrict nonconductive material from entering the work area, that material should be isolated from sensitive products.

Mobile Work Positions (Packing, Mass Soldering, Etc.)

At standing work positions, conductive footwear may be worn by personnel if wrist straps would impair required mobility or are a safety hazard. However, personnel using conductive footwear must stand on the conductive or dissipative flooring or conductive tile to be properly grounded.

Transporting Products

To avoid direct hand contact, products should be transported in static-safe containers that include but are not limited to those specified in Table 7-1 (tubes, bags, boxes, etc.).

> ***Note:*** An in-depth discussion of the design concepts for these containers and their use is presented in Chapter 12, "Packaging Considerations."

Carts

Carts should be designed so that there is no direct contact between the conductive materials of the carts and the ESD-sensitive product. For instance, carts should be designed to prevent contact between the conductors of a PWB assembly and the cart (that is, the slots that guide the plug-ins and provide clearance between them). However, products in approved transport media can be placed on the conductive surface of carts.

In Class 0 and Class I shops, carts must be equipped so that they can be grounded during loading or unloading if bare product might be contacted. In areas with conductive or dissipative flooring and finishes, carts should also be grounded during transit (that is, chains or conductive wheels to conductive flooring or floor finish) as an added precaution.

> ***Discussion:*** It is important to ensure isolation between the conductors of the sensitive product and carts. This is because carts are highly prone to charging while in motion. Therefore, it is desirable to prevent charging as well as the possibility of direct discharge from the cart to the sensitive product. Typically, once the

carts have reached their destination, grounded personnel position them and thereby remove most of the charge from the cart without damaging the contents. This is similar to what happens while handling PWB assemblies in tote trays and is adequate protection for the less sensitive products in Classes II and III. However, in the more sensitive classes, it is necessary to ensure complete charge removal during the loading and unloading process by adding a positive ground to the carts. Strictly as an added precaution, it is recommended that all carts be equipped with a drag chain to reduce the degree of charging while in transit. Twelve to eighteen inches of contact with the floor is recommended; therefore, the chain will need to be attached at both ends to avoid creating a tripping hazard.

Static-Safe Packaging

All products containing ESD-sensitive devices must be shipped in static-safe packaging that is marked with an industry-standard Static Awareness Symbol. (See Figure 7-1.) The lettering is black on a yellow background. All of the packaging materials used for these items must be static-safe. The static awareness label identifies contents that are static-sensitive and require special handling precautions at the receiving end. Static-safe material test specifications are defined in EIA Standard No. 541, "Packaging Materials Standards for Protection of Electrostatic Discharge Sensitive Devices," from the Electronic Industries Association.

Note: Some tape-and-reel packaging does not comply with packaging requirements and should be only used with due caution and proper qualification. It is not recommended for use with Class 0 and Class I devices.

Figure 7-1. Static awareness symbol

Static-Safe Bags, Shipping Tubes, and Tote Boxes

Employees must load and unload sensitive products in the manner determined by the shop ESD classification. In Class 0 and Class I shops, totes, or other containers, and carts (conductive or ESD-safe) must be properly grounded when being loaded or unloaded. In Class 0, I, II, and III shops, when unloading or loading metal trays, racks, or other metal containers that may contact the conductors of sensitive products, the containers must also be connected to ground through an ESD-safe surface or some other ground path. IC shipping tubes must always be grounded during loading or unloading.

> ***Note:*** At this time, IC shipping tubes are not recommended for Class 0 devices. This is because of recent reports that most tubes charge devices sufficiently to jeopardize Class 0 devices.

These requirements apply only when handling sensitive products directly. In other words, the containers may be transported or handled without grounding as long as the contents are not directly contacted by personnel.

> ***Discussion:*** All tubes that are used for shipping and handling integrated circuits must be grounded at all times during either the loading or unloading process. This is necessary to prevent any charge transfer from the IC shipping tubes to the devices upon exit. A device charged in this manner is subject to charged-device model failure.
>
> In fact, all IC shipping tubes are inherently prone to charging devices to some degree and are therefore not recommended for the ultrasensitive devices in Class 0 applications. Tests have shown that most IC shipping tubes will charge devices sufficiently to jeopardize those with thresholds below 200 volts. This precaution is subject to change as better IC shipping tube construction materials become available and should be reviewed periodically.
>
> As stated in Table 7-1, dissipative tote trays are required in Class 0 and Class I areas only. However, when nondissipative tote trays need replacement, they should be replaced with dissipative tote trays. This provides a realistic and cost-effective transition period that will ultimately result in a fully-equipped manufacturing facility.

The grounding of tote trays is required in Class 0 and Class I areas due to the highly sensitive nature of the devices being handled. This grounding can be achieved by placing the trays on an approved dissipative surface. In less sensitive applications, it is a reasonable risk to rely on the inadvertent grounding that results from grounded personnel handling the tote trays.

The design of the tray could help to ensure this type of grounding by including a lid which would prevent direct contact with sensitive devices without first touching the tray and removing the charge. Furthermore, if the tray holds the circuit boards in slots perpendicular to the outer surfaces, the effects of the coupling coefficient to external fields will be more forgiving. (See Chapter 12 for more details.)

Dissipative Table Mats

The primary purpose of dissipative (limited to 10^5 through 10^9 ohms per square {***see Discussion below***}) table mats is to ground devices and bench fixtures, not personnel. The surface of the mat should maintain their resistance under all humidity conditions and acts to dissipate surface charge. Dissipative mats are typically hard grounded. Parts, bags, and tote boxes, as well as people contacting the mats, will not build up static charge. Factory standards should specify installation procedures, including proper grounding methods.

Discussion: Dissipative table mats and tabletops are required only in Class 0, I, and II areas and are not required in Class III or IV areas. This requirement is based on experience as well as the data presented in Chapter 4, "An Economic Analysis."

For instance, a case study presented in Chapter 4 revealed that devices with a 350-volt sensitivity (Class I) were protected by using nothing more than wrist straps and dissipative tote trays. They were handled under normal manufacturing conditions without using dissipative tabletops. Although no ESD damage was detected in the protected population, as much as 50 percent of the unprotected population was damaged by ESD.

While it might be concluded that specifying a dissipative tabletop for Class II devices is more stringent than was indicated experimentally, it is consistent with the premise of providing added layers of protection to compensate for human error or equipment malfunction.

The dissipative range for table mats and tabletops has been limited to 10^5 to 10^9 ohms per square based on Bell Laboratories research. This range ensures that charges can be removed at a safe rate, neither too fast nor too slow. If a work surface (such as stainless steel) is too conductive (less than 10^5 ohms per square), a charged device can be damaged upon contact per the charged-device model. Conversely, if the work surface is insufficiently conductive (greater than 10^9 ohms per square), the charge may not be removed completely or swiftly enough to prevent damage by a grounded employee.

Dissipative Tabletops

The preferred material for tabletops is dissipative (limited to 10^5 through 10^9 ohms per square), which serves the same purpose as the dissipative table mats—that of grounding devices and bench fixtures.

Caution: See the subsection earlier in this chapter on conductive materials. Generally, a laminate-type, hard material is used to dissipate a static charge and the conductive buried layer (see Appendix 3) must always be grounded. These tops are extremely durable and should not require replacement under normal usage. Factory standards should specify installation procedures including proper grounding methods.

Dissipative Floors/Floor Finishes

Dissipative floors and floor finishes must be used with approved ESD foot protection to effectively ground personnel. They also provide a means of grounding static-safe carts.

Flooring materials and finishes must be approved by the local ESD coordinators in order to qualify as static-safe. Auditing procedures for monitoring their ESD properties and maintenance must be instituted.

Discussion: Dissipative floor finishes are required in Class 0 areas as an added precaution and are recommended in all areas where grounding for mobile employees is needed. Conductive finishes not only add a conductive surface to the floor, but they also ***significantly reduce the degree of charging*** that takes place. As a result, they build in an added layer of protection as a safeguard against the possibility of human error or equipment failure. These floor finishes often replace the need for conductive or dissipative mats.

Conductive Edge Connector Shunts

Using a conductive edge connector protector or shunt reduces any static potential differences that may develop between conductor paths on PWB assemblies. It also eliminates the risk associated with the nonconductive edge connector protectors. It is required on Class 0, I. and II PWB assemblies.

Discussion: Conductive plastic shunts can excessively charge a PWB assembly upon removal. In fact, the residual charge often exceeds an acceptable limit of 0.2 nanocoulombs per square inch. Therefore, be sure to evaluate the triboelectric properties of shunts before selecting one for use. An antistatic dissipative shunt is a reasonable alternative for most applications, but may not be adequate for Class 0 applications.

Extraordinary Measures

Devices with thresholds below 200 volts (Class 0) are considered to be extremely sensitive and will require extraordinary handling measures to ensure reliable protection. In fact, devices rated below 100 volts may be impossible to manufacture reliably. The extent of the precautions needed is not fully understood at this time; therefore, ***it is advisable to obtain the recommendations of a corporate consultant***. If at all possible, the design should be changed to specify a Class II or Class III component. Recognizing that this is not always possible, the following discussion may be of some assistance.

Specific precautions need to be tailored to each situation, and some techniques could preclude the need for others. For instance, an in-process shunt across the leads of a sensitive device may be sufficient. However, there are recorded instances of devices that have failed even under these conditions. Other measures include those listed elsewhere (see the following subsection,"Other Controls"), as well as the use of redundant wrist straps, continuous monitoring of those wrist straps, and conductive chairs. The use of IC shipping tubes for Class 0 devices is not recommended at this time. This is due to recent reports indicating that most IC shipping tubes can charge devices sufficiently to jeopardize Class 0 devices.

Discussion: In dealing with these ultrasensitive devices, it has become apparent that cookbook methods and solutions are generally insufficient. This is especially true when the damage occurs as a result of the automated manufacturing facilities. Therefore, it becomes important to tailor the solutions available based on the advice of a corporate or industry consultant to establish a cost-effective means of dealing with devices at Class 0 thresholds. As stated in Chapter 4, "An Economic Analysis," this approach reduced AT&T's operating costs by $6 million, in one instance, and yet the added costs for implementing the ESD precautions amounted to less than $1000. This was done by protecting the device and not trying to purify the whole room. This is a fundamental departure from cookbook methods and has proven to be far more effective for ultrasensitive applications.

Other Controls

Specific applications may require more stringent or more specialized control techniques than the ones specified in this Handbook. The engineering organization will have to prescribe special techniques in unusual situations. For example, people working with extremely sensitive parts may be required to wear antistatic smocks or work in a controlled humidity environment. Ionized air generators (which neutralize charges on surfaces) may be used at workstations, in areas where hazards cannot be eliminated because of process requirements, or when highly-sensitive devices (Class 0 and Class I) require zero charging. Conductive foam may often be used for packaging and handling these highly ESD-sensitive devices. Conductive cart wheels and dissipative floors may also be required.

Transporter and other conveyorized shops are potential troublemakers unless antistatic belts and metal roller bushings are used. Gravity-fed flow racks with metal skate wheels should be installed. In addition, all mechanized equipment used for moving material must be wired to ground, and the proper materials must be selected to avoid triboelectric charging. Soldering irons and desoldering tools are also potential hazards. In some instances, soldering iron tips are not grounded and may produce transients. DOD Standard 2000 provides guidelines for control of soldering tips. Solder suckers that are not static-safe should never be allowed at the workplace.

Applying or removing adhesive tape presents significant risk. It is permitted (but ***not recommended***) for ESD-sensitive items. It must be applied in the presence of air ionization or some other means of charge neutralization.

Factory Standards

Local factory standards should be established by engineering in compliance with the specified ESD control handbook.

Discussion: The documentation and consistent application of a factory standard for installing control facilities, such as workbenches, wrist strap grounding provisions, or floor mats, is a vitally important aspect of the ESD program. It is through the formal documentation of factory standards that facilities can be consistently and safely installed or upgraded. These facilities are an integral part of the program and have a direct bearing on its overall success. A failure to recognize this can lead to safety problems, unnecessary device failures, unmanageable training difficulties, and a loss of credibility for the entire program. Therefore, it is strongly recommended that a well-planned factory standard be developed and followed closely.

It is unlikely that one set of factory standards will be suitable for all manufacturing locations within a large corporation. Each location will need to develop standards suited to its own situation. These standards should be based on a common understanding of safety and grounding requirements. A corporate ESD committee is a good forum in which to establish the general guidelines.

Points To Remember

- The requirements in this chapter form a consensus standard (AT&T *Electrostatic Discharge Control Handbook*) developed by and for use at all AT&T manufacturing locations. They are based on the handbook developed at the North Andover facility.

- These ESD control techniques provide a common answer and a realistic solution for the electronics industry.

- Only realistic requirements can be enforced, and they must be formally documented.

- The requirements must be based on a sound technological understanding of the fundamentals of electrostatics, including the failure models.

- When engineering takes into account technology, training needs, and human nature, the result is a program of realistic techniques that are both manageable and effective.

- Redundancy, without excess, provides needed layers of added protection to ensure that sufficient safeguards are in place and that human error will be minimized.

- The requirements must vary with device sensitivity in order to be both cost-effective and realistic. Ultrasensitive devices clearly require far more protection than is economical or practicable for less sensitive devices.

- The variation in requirements can be accomplished without requiring employees to learn several different techniques. With few exceptions, the ESD control system can be designed to offer better protection for more sensitive devices and still have little impact on the overall depth and scope of training.

Chapter 8

Implementing an Auditing Program

On most mornings the ESD auditing inspector begins the workday at the Merrimack Valley Plant by entering the previous day's data into the computer. Then he proceeds to the manufacturing floor and continues the data collecting process for the current auditing cycle.

Equipped with a cart of test equipment, an inspector's guide, and a trained eye, he audits two shops each day. At each shop he inspects a predetermined number of randomly selected workstations for deviations from the prescribed ESD control procedures. He makes visual checks and electrical tests. Each noted deviation is recorded in a one-hundred plus item checklist. Written notes are added in a comment section. After completing his work in a shop, he moves on to another and repeats the data-gathering process.

The data-gathering process continues on a daily basis, except when the inspector is required to perform a special training assignment or an investigative project. After all of the departments have been audited, the cycle ends and the summary reports are printed. Following the distribution of these reports to management, the next cycle begins.

The reports provide graphic summaries of the relative effectiveness of the ESD control program. By analyzing the reports, comparisons can be made between succeeding auditing cycles to look for trends. While some trends reveal successes, others may reveal possible problems. Comparisons can also be made between departments or the various types of deviations for a given auditing period. The graphs might show that the most serious plantwide problem is faulty wrist straps or that personnel in Department A need further training in the use of heelstraps. It is the comparisons between departments and between types of deviations that identify for the coordinator the most serious problems.

This chapter explains how to implement the data-gathering program. While the reports are very powerful and far reaching, the program for gathering the data is relatively simple. The next chapter explains how to read and use the auditing reports. With a steady flow of auditing reports and the ability to use them, the coordinator can manage the program wisely and scientifically. In the hands of a skillful coordinator, the information gained from auditing can simultaneously serve to bind the individual parts into an integrated whole program and to drive the entire program towards continual improvement.

The auditing process is the binding force behind the entire ESD control program.

Why Auditing Puts the Coordinator in Charge

With a technique for identifying the program's most serious problems (Figure 8-1) and the ability to spot and report trends (Figure 8-2), the coordinator has the ultimate tools necessary to be in charge. ***The program can then be managed on a scientific basis***. Decisions about how best to run the program will be made based on the available data. For example, the coordinator can pull together all available resources to solve a heelstrap problem or to schedule specific additional training for some employees. He can modify the qualification procedure for certain equipment, or ask a department manager to prepare a plan of corrective action.

Without auditing reports, the coordinator does not have essential, timely information necessary to identify and solve problems. He has no choice but to manage from one crisis to the next or, even worse, assume that there are no problems and not manage at all. There have been companies where once the equipment was purchased and installed and the employees trained, management assumed that the ESD problem had been solved and made the costly error of moving the coordinator to a new assignment.

Accepting the notion that there are many problems in any ESD control program should not be difficult especially when one considers the number of items that could be defective and the number of employees who must master new work behaviors.

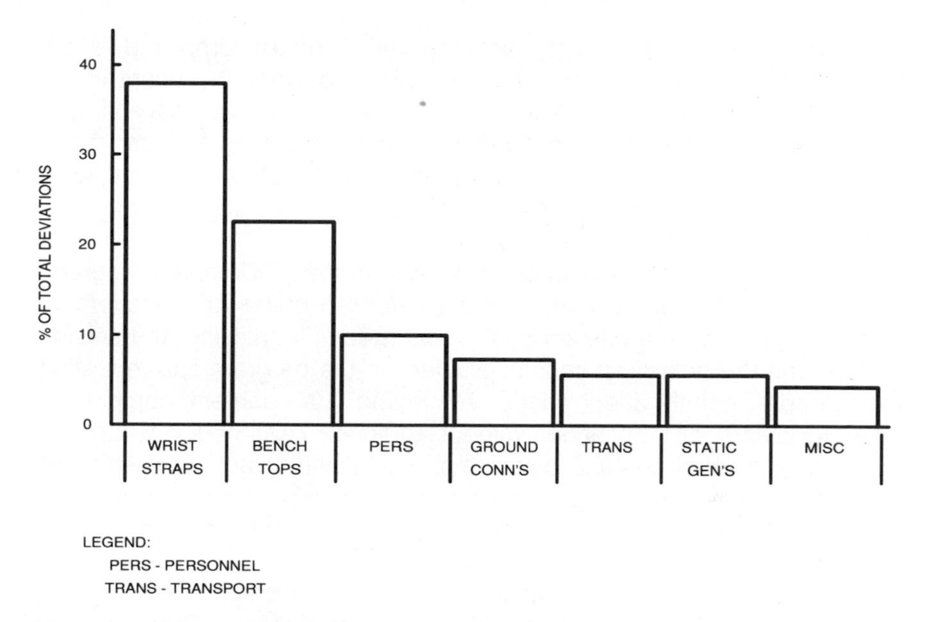

Figure 8-1. A typical Pareto chart used to identify and prioritize ESD control problems

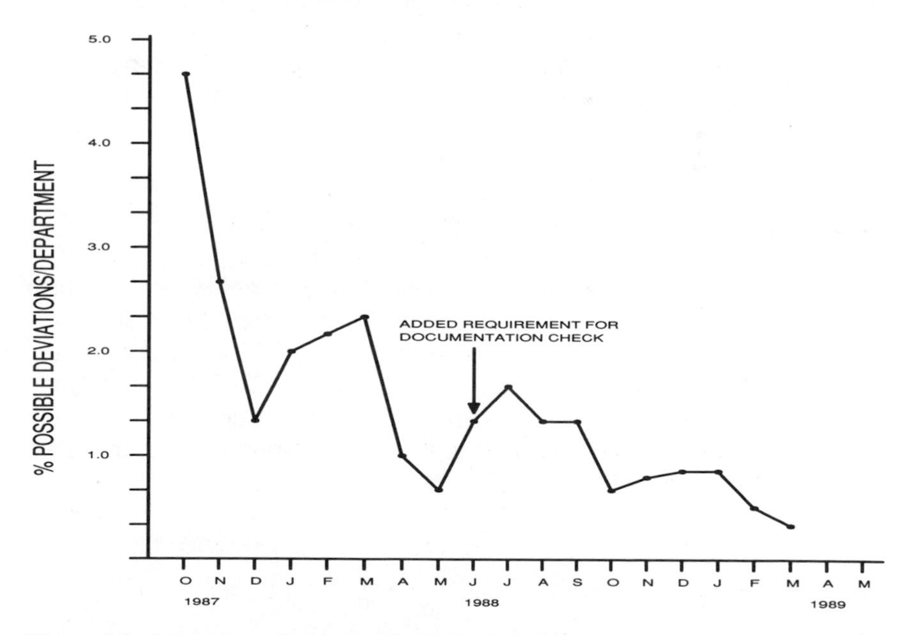

Figure 8-2. A long-term trend (trend chart) of deviations recorded at one of AT&T's manufacturing locations (AT&T Reading Works)

All available data supports this assumption. Studies completed at our North Andover plant in 1985 revealed that more than 50 percent of the ESD control equipment in use was either defective or inoperative. These figures were confirmed by a 1988 industrywide study of thirty major electronic companies.[1] Astoundingly high defect levels were found: wrist straps, 35 percent; heelstraps, 80 percent; conductive flooring, 75 percent; and IC shipping tubes, 40 percent.

In fact, every audit will highlight problems in the ESD control program. That is why the ESD coordinator must establish continual improvement of the program as the primary goal. This means identifying and solving one-by-one the next most serious problem in the program and repeating the process month after month. Achieving 100 percent compliance takes time and can not happen without auditing. Examples in Chapter 9 concerning wrist straps and bench tops will illustrate how problems still existed in a fairly mature ESD control program and how solving those problems helped the program, on a plantwide basis, approach zero deviations.

Overall, by identifying the most serious problems uncovered by the audit, one can begin a sequence of problem solving. Problems get solved one at a time, and then the next set of major problems are identified and solved. This sequence of problem solving produces documented, steady, continual improvement. Reporting the progress to management keeps them informed and helps to sustain their commitment to the program.

What Must Be Audited

The inspector audits how well ESD control equipment works and how well employees comply with ESD control procedures. In addition, the inspector collects data on a third aspect of the ESD control program: How well ESD control is monitored in the shop. At AT&T, this monitoring effort called process checking, is part of an effort to build quality into the manufacturing process rather than to test the finished product and then try to correct a problem.

Process Checkers

Process checkers work for shop managers, who have the ultimate responsibility for ***ensuring day-to-day compliance*** with quality control procedures within the manufacturing process. The process checkers, who are trained and are given a detailed written guide to follow, check all existing procedures and methodologies. For example, silicon integrated circuit bonding is a critical manufacturing process and is monitored by process checkers every 15 minutes. ESD control is also a manufacturing process that requires similar quality control. The method used by process checkers is the traditional quality control procedure of collecting data and plotting control charts which are then used to manage the process.

The ESD process checker performs both visual inspections and electrical tests and plots all defects in a daily control chart (Figure 8-3). He/she checks the general area and tests the following workstation items: wrist strap system, wrist strap only, heelstrap, grounding receptacle, tabletop, table mat, table ground cord, floor mat, floor mat ground cord, and microscopes. Some items, such as wrist straps, are electrically tested once per day for each shift. Other items, such as microscopes, are tested once per week. Almost all items are checked visually on each process check.

The written guide gives the process checker specific procedures for dealing with each deviation. When the process checker discovers a deviation, there are prescribed corrective actions that must be taken immediately. For example, if the checker observes an employee not complying with procedures, the problem is then reported to the appropriate supervisor. On the other hand, if he/she finds a defective wrist strap, it is removed from the workplace immediately.

Process checkers work for manufacturing supervisors, not ESD coordinators. However, the ESD coordinator might suggest the following job qualifications for process-checking ESD control. The employee will:

- Be entrance-level or slightly higher.
- Have passed a training program that includes basic certification, process checking, and charting procedures.
- Have personality traits that include being steadfast, nonabrasive, diplomatic, assertive, and forthright.

DAILY CONTROL CHART

DAILY SUMMARY — ESD DEVIATIONS-VISUAL — AREA *Circuit Pack Assembly*

section chief initials

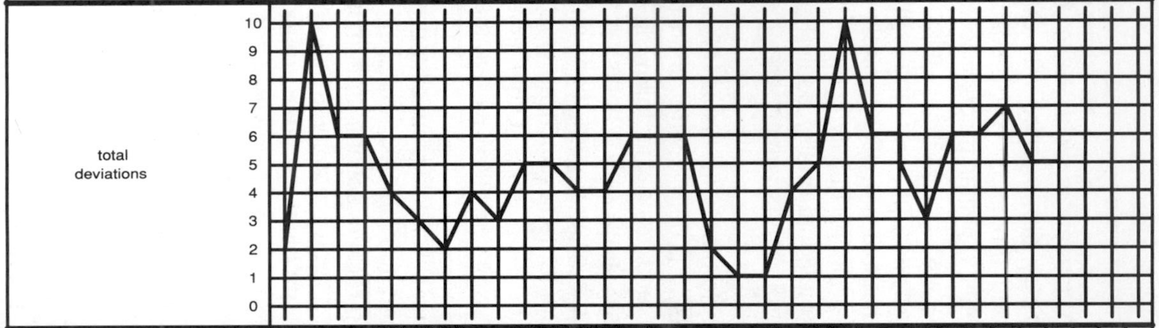

wrist straps		1		2	2		1	1	2	2	1			1	1	1	1		1	1				1		1							
improper handling of trays		6	2	1	2	2	1	2			1			2		2											1			1			
unsafe surfaces		1	1			1			1	3	1	3	3		4	3	1	1		1	5	10	6	5	3	5	5	7	3	3			
static-generating material	1	2	3	2							2	1	1	1	1	1				2													
other (note on back)	1			1				1						2	0																		
total defects	2	10	6	6	4	3	2	4	3	5	5	4	4	6	6	6	2	1	1	4	5	10	6	6	3	6	6	7	3	4			
total # of items checked	240	239	241	241	241	241	240	241	241	241	241	241	241	241	242	242	242	242	242	242	242	242	242	242	242	242	242	242	242	242			
date	8-31	9-1	9-5	9-6	9-7	9-8	9-11	9-12	9-13	9-14	9-15	9-18	9-19	9-20	9-25	9-26	9-27	9-28	9-29	10-2	10-3	10-4	10-5	10-6	10-10	10-11	10-12	10-13	10-16	10-17			
check #	1	2	3	4	5	6	7	8	9	10	11	12	13	14	15	16	17	18	19	20	21	22	23	24	25	26	27	28	29	30	31	32	33

Record defect details on back of chart. Reference to check #.

Figure 8-3. A daily control chart used by process checkers to record and plot ESD deviations

- Understand the shop and the manufacturing processes.
- Be a proven good employee who is responsible and accountable and shows pride in work accomplished.
- Show promise of being a local expert.
- Be able to communicate well with other employees about quality control.

A process checker with these qualifications will be effective in controlling the ESD process. However, there should be only one person per department with the responsibility for ESD process checking. This task may or may not be a full-time assignment, depending on the size of the department. Regardless, limiting the responsibility for ESD process checking to one person is a critical factor in the communication process among the ESD coordinator, the plant inspector, the ESD process checker, and manufacturing personnel. It is perhaps the only way that a large department can achieve perfect compliance or zero deviations.

How to Obtain Meaningful Data

Sound conclusions that lead to useful recommendations made with confidence are only possible with accurate and reliable data. The old adage of "garbage in, garbage out" applies as much to ESD auditing reports as it does to any other computerized analysis of scientific data. Also, the data-gathering effort must be frequent enough so that trends can be noted and serious problems uncovered and resolved before too much damage is done. This section tells how to obtain ***auditing data that is reliable and useful*** and what conditions are needed to obtain consistent data at an acceptable frequency.

All scientific studies require a common unit of measure. Without such a measure, valid comparisons are not possible. A procedure must be in place so that any measurements made at any place and at any time can be compared with all other similar measurements. Physics begins with a meter and a second. Electricity uses a volt. Astronomy uses light years. ***Auditing ESD control to measure compliance uses deviations-from-prescribed-procedure***. Examples of deviations include a broken wrist strap wire, a defective wrist strap, or an ungrounded work surface. With an inclusive list of all possible deviations, observations made at workstations should be comparable.

The number, or frequency, of deviations (such as 20 wrist strap deviations in Department A and 12 wrist strap deviations in Department B) is a relative measure of compliance. The number of deviations is plotted on the Y-axis (the dependent variable) on all graphs. An exception is when a derived figure, the percent of total deviations, is substituted. This information permits the comparison of many types of independent variables, or attributes data, such as wrist straps and bench tops, or employee performance between departments.

Since the number of deviations is a quantifiable measure, the information can be entered into a computer easily. Using customized software, the data can be manipulated to make many different types of comparisons and can be printed as either line or bar graphs. Using a computer in addition to expanding the analytical power makes recordkeeping a less labor intensive task.

Thus, to obtain consistent data, deviations-from-prescribed-procedure should be used as a basic unit of measure. Then, to ensure that all observations lie within a defined set, the number of different types of deviations should be predefined by preparing a checklist. Lastly, an inspector's guide should be written to describe how tests and observations are to be performed. These three factors, a common unit of measure, a checklist, and a test procedure, will help process checkers to obtain consistent data.

A Full-Time Inspector

When our program started, each department had a part-time auditing inspector who was a member of the quality organization. It was not uncommon for the same deviation to be reported differently from department to department. For instance, suppose three inspectors see a row of five stand-up workstations with one floor mat that runs the entire length. The floor mat is not grounded, and two employees are standing on it. Inspector A records two deviations. He sees two ungrounded people standing on the mat. Inspector B looks at the same situation and records five deviations, five workstations that are not grounded. Inspector C records one deviation, a single floor mat that is not grounded.

To eliminate this problem and build ***greater consistency*** into our data, we appointed one well-trained, well-qualified, full-time auditing inspector to cover the entire plant who was also a member of the quality organization. With one inspector, it doesn't really matter whether he/she sees one, two, or five deviations in the above example, provided that he is consistent. Thus, whenever he sees people at workstations standing on a long ungrounded floor mat, the number of deviations recorded will be consistent and allow the coordinator to make valid comparisons. However, can one person really audit an entire plant?

The answer to this question is a qualified yes. One inspector can audit an entire program providing two conditions are met. First, process checkers, or their equivalent, are needed to provide the day-to-day supervision of the program. Second, by using statistical sampling techniques, one inspector can collect data for an entire plant. Visiting only 20 workstations for each department gives perfectly valid results. Not only can one person audit an entire plant, ensuring consistent observations, but a single competent inspector, accepted by the employees, can help in the training aspect of the program by giving tactful corrective suggestions.

One inspector should be able to complete eight to twelve auditing cycles each year. Ideally, the auditing reports should be ready on a monthly basis to detect trends and to spot radical changes in the program. This means that ideally the inspector should complete an auditing cycle by visiting every department in the plant every month.

Auditing Reports Supersede Yield and FMA Data

When a steady flow of auditing reports start to arrive on the ESD coordinator's desk, yield reports and failure analysis studies become supplemental tools for detecting and analyzing ESD control problems and failure analysis should never be used as the primary means of maintaining daily compliance with proper procedures. With yield reports and failure analysis, there is too much of a lapse between the time that a problem occurs and when it is finally discovered. However, the ESD coordinator should continue to study available yield information. A precipitous drop in yield could mean that the problem was ESD-caused by means of automated processes.

The Auditing Inspector: Qualifications and Duties

The auditing inspector's first responsibility is to collect data for the auditing reports. The next major responsibility is to help train the work force. While making rounds of the entire plant, performing visual inspections and electrical tests, as well as taking notes, conversations provide on-the-job training to employees and process checkers. Some training is intentional; while at other times, it is an incidental part of the many conversations between coworkers and the inspector, who is accepted as an authority on ESD control.

Qualifications

Qualified candidates for ESD auditing inspector do not need a technical education. They should be able to learn the technology of ESD control and how to detect deviations. They should have an understanding of the manufacturing process and organization of the factory. The candidates should be able to downplay being an inspector. Also, they should be able to teach others how to do their job better. Last of all, they must be able to handle confrontation calmly.

Special personality traits, including being steadfast, nonabrasive, diplomatic, assertive, and forthright, are important. The inspector needs to make decisions in a forthright manner and to defend them with diplomacy. He/she must walk a tightrope. The entire program rests on the accuracy of his/her observations and reports.

However, there must be room for differences of opinion. The inspector could be wrong and must recognize that a dispute conducted with courtesy and respect can be informative to both parties. He/she should be able to point out small day-to-day problems, and then patiently show employees how to do their job correctly. He/she should not consider himself/herself an examiner or a policeman. Rather, he/she should collect data in an objective fashion and educate others in a warm, helpful manner. In short, he/she should be able to communicate with other employees in a constructive manner.

He/she should be able to learn the technology of ESD control in great depth. Most of the inspector's training will come from the ESD coordinator and will often require up to three months of intensive on-the-job training. However, candidates will be required to have previously passed all of the available ESD courses described in Chapter 11, "Training for Measurable Goals."

They should know the plant layout well and be familiar with most of the products. This knowledge will help them choose daily assignments and to recognize when a significant change has taken place. This is especially important when new products are introduced that require ESD controls or more stringent requirements. For example, a new product containing a more sensitive device might affect what ESD equipment should be used by changing the sensitivity class.

Duties

The following description of an inspector's duties will help in understanding the subtle aspects of the skills and personal attributes needed to be a successful auditing inspector.

As mentioned briefly at the beginning of this chapter, the inspector's (Figure 8-4) daily duties at the Merrimack Valley Plant begin by entering the previous day's data into the computer. His entries include coded items from his checklist and written comments. He takes special care in phrasing his comments so that they are constructive and helpful for shop management. Next he reviews the shops that remain to be inspected for the current auditing cycle and chooses two sites for that day's visits.

After gathering the test cart (Figure 8-4), equipment, and supplies (Table 8-1), he heads out to his first test site of the day. On the way, since he knows people throughout the plant, he exchanges greetings with a number of people and, in one instance, stops and discusses a question about testing bench tops. Before entering the shop to be inspected, he determines which 20 positions will be checked. This decision is made as objectively as possible.

Figure 8-4. The ESD inspector and test cart

Table 8-1. The Inspector's Test Equipment, Supplies, and Documentation

Equipment and Supplies:

- Surface resistivity meter
- Resistance to ground meter
- Wrist strap tester and ground plate
- Wrist strap calibrator
- Electrostatic locator
- Materials for tribocharging
- Multimeter (vom)
- NFPA probes
- Megohm meter
- Ion current meter
- 6-outlet extension cord
- Horsehair brush
- Out-of-service signs
- Grounding hardware for benches
- New wrist straps for replacements
- New heelstraps for replacements

Documentation Packet:

- Inspector's Manual with checklists
- ESD Control Handbook
- Process Checking Instruction
- ESD Bulletins
- ESD Policy Statement

He enters the shop unannounced, seeks out the ESD supervisor, and explains that he will be doing an audit. Supervisors, process checkers, and engineers are invited to follow him and talk about the tests that he is performing or about the ESD control program in general. These conversations keep the communication channels open while serving as part of the on-the-job training effort.

After visiting two workstations, he realizes that the process checker is new and is not properly trained. The deviations related to insufficient training are recorded. He provides some on-the-job training as he explains the deviations and recommends that the process checker receive additional training from the training organization.

At one point in his visit, the supervisor challenges the inspector's observation. The inspector observed that due to insufficient training the wrist straps had not been tested according to prescribed procedures. With great diplomacy, he explains his observations and then calls in the plant ESD coordinator for help. After all of the different points of view have been exchanged for over five minutes, all parties agree that the inspector's observation is correct. Satisfied with the outcome, the shop personnel agree to take immediate corrective action.

On his rounds of the shop, the inspector notes that many employees do not know how to test wrist straps. He records only one deviation, deciding (after obtaining approval from the shop supervisor) to conduct a brief training session on testing wrist straps.

The inspector takes all defective wrist straps and equipment and puts them in a box on his cart. These items will be analyzed by the ESD coordinator.

The inspector and the process checker do many of the same requalifying tests; however, the inspector has a number of added responsibilities. Thus, as the inspector tests the equipment, he also checks the process checker's work and discusses correct procedures. The inspector also requalifies bench tops, tote trays, floor mats and floor finishes, items not tested by the process checker.

The department is held only accountable for the equipment that is listed in the process checker's instruction documentation and that can be tested with a wrist strap tester. Other items that fail requalification are only noted in the auditor's reports. However, all information on defective equipment is used to review purchasing procedures, to notify vendors about the reliability of their equipment, or to initiate corrective actions.

After the inspector completes the shop audit, which was achieved in three hours, he travels to the next shop he has chosen for testing on that day. At the second shop, he repeats the data collection process and on-the-job training process.

The information on the checklist and report form will be entered into the computer, and the shop will receive a copy of the preliminary report in 24 hours. Shop personnel are encouraged to study the preliminary report and to contest any items that they feel are incorrect.

If a shop supervisor chooses to dispute an item, he can; but he must address his concerns to the ESD Coordinator, not the inspector. The ESD Coordinator is the only one with the authority to change any item recorded by the inspector. These discussions bridge an important communication link, and a lot is learned about ESD control during these conversations. In most instances, the inspector's observations are correct, but once in a while the coordinator will need to either overrule the inspector or to change a requirement.

On occasion, the inspector is pulled away from his primary responsibility of collecting data and doing on-the-job training to work on a special assignment. These special assignments often last a day, or at most a week. For example, our inspector has audited suppliers and subcontractors and has helped with updating manuals. Also, after the inspector has been doing the job for a couple of years, he becomes a knowledgeable resource who can often help solve an unexpected crisis. In one instance, he was assigned to work on a bench top problem. This project took three months and without his help would have been very difficult to solve.

The Inspector's Guide and Checklist

The Inspector's Guide used at AT&T is about 50 pages long, and the accompanying checklist has over 100 items grouped into the 17 major categories listed below:

wrist straps	ground receptacles	footwear
soft table mats	hard table mats	floor mats
conductive flooring	transport media	storage
personnel compliance	signs and class	ESD floor finishes
packaging	wrist strap checkers	air ionizers
process checks	static-generating materials	

To write a guide and checklist for your plant's ESD inspection procedure, refer to your ESD control handbook. It should explain the program as well as define techniques for 100 percent compliance with ESD control. This information is the basis for the Inspector's Guide. Therefore, study the handbook, look for observable pass-fail items that point out compliance or noncompliance, and create a checklist similar to the one in Figure 8-5. Then repeat the process for each of the 17 major categories.

The completed checklist can be converted to an inspector's guide by expanding the list to include specific procedures (Figure 8-6). These procedures should describe, in detail how all of the tests and observations are to be made on a pass-fail basis.

ITEM	QUANTITY	DESCRIPTION
Wrist strap	----- 1.1 (D)(P)	Wrist strap cord resistance, open or intermittent
	----- 1.2 (N)(P)	Wrist strap snap condition
	----- 1.3 (N)(P)	Wrist strap condition
	----- 1.4 (N)(P)	Wrist strap banana plug condition
	----- 1.5 (D)(P)	Wrist strap process charting evidence not available
	----- 1.6 (N)(P)	Intermittent cord cannot be verified
Wrist Strap grounding	----- 2.1 (D)(F)	Wrist strap grounding receptacle not present

Notes:

(1) D = deviation N = notes P = process F = facility

(2) A deviation (D) is a major item that must be corrected immediately and is included in all final reports. A note (N) is less serious and is an indication of pending problems. A process (P) is related to human behavior, where as a facility (F) relates to control equipment.

Figure 8-5. A portion of the inspector's checklist, which is used to record deviations and to enter auditing results into the software data base

II. Specific Plan Requirements

1. Wrist Straps

1.1 Wrist Strap Cord Resistance (D)(P)

The ESD wrist straps cord should be measured with a voltmeter (VOM) that has a minimum precision of 50 kilohms per volt. The 100x resistance scale should be used, and the reading should be greater than 750 kilohms and less than 1.2 megohms. The cord should be stressed from side to side and subjected to a pulling stress to discover intermittent conditions. The wrist strap tester should be used to test the bracelet and cord system integrity. If the tester does not verify the intermittent condition, it should be entered in the Shop Report as a "note" item by using Subsection 1.6 of the ESD Inspection Guide.

1.2. Wrist Strap Bracelet Snap (N)(P)

The snap should be mechanically sound. When plugged into the wrist strap, it should hold against a force of 1 to 5 pounds as applied against the cord fastener at the snap extension.

1.3. Wrist Strap Bracelet Condition (N)(P)

The bracelet should not be frayed or show other evidence of undue wear. The bracelet elasticity should be such that it fits the user's wrist snuggly and does not allow the bracelet to roll back.

1.4. Wrist Strap Banana Plug Condition (N)(P)

The blades of the banana plug should not be bent or dented, nor should the plating be worn away to the base metal. The plug should hold in its socket under normal use and against a force of 1 to 5 pounds.

Figure 8-6. A portion of the Inspector's Guide

ESD Data Base Management Software

The data gathered by the inspector must be readily available for analysis and` maintained indefinitely. Therefore, a data base management software system tailored to the needs of the ESD auditing process is a necessary tool.

The ***software needs to be user friendly*** and menu driven so that the inspector can maintain the data base and generate the appropriate graphics outputs easily. It must also be flexible enough to allow for the inevitable organizational changes that all companies undergo. When reorganizations occur, the history of all departments must be preserved and the organization numbers must be changed to coincide with the new numbers. Distribution lists will change and occasionally recorded deviations will be modified or deleted.

Entering the data can be simplified by assigning item numbers to the various deviations. The inspector can enter one number for the deviation, another for the quantity, and others for the department number, shift, date, and sample size. (Figure 8-7). This will automatically generate consistent descriptions in the reports and keep the inspector's data entry time to a minimum. This consistency is an invaluable aspect of the communication process.

The software can then, upon request, generate preliminary reports, final reports, and an assortment of graphs. The preliminary reports should automatically include a cover page with a distribution list, the type and quantity of deviations, the inspector's comments, and the appropriate sections of the Inspector's Guide.

```
                ESD - Shop Inspection Results

Dept: 55555  Shift(1,2, or 3): 1  Date (YYMMDD) 891110
                Sample Size: 20

                PWB Assembly Department

ITEM QTY TYPE DESCRIPTION
 9.3  1    D   Product stored on unsafe surface
17.2  1    D   Improper electrical test methods
```

Figure 8-7. Input screen for the ESD data base management software. The inspector's entries are in boldface type.

Deviations reported in preliminary reports are subject to change, and therefore the software must be able to accommodate this easily.

The final report should include most of the preliminary report, a cover letter from the ESD coordinator, trend charts for each department, by shift, and a Pareto chart and a trend chart for the entire plant. The inspector's comments and sections of the Inspector's Guide can be eliminated from the report, at this point, but should be maintained in the data base for future reference.

The graphics generated by the software are illustrated in the next chapter and need to be both extensive and flexible. They provide the coordinator with a wide variety of options that make it possible to analyze all aspects of the ESD program. This ESD program analysis will advance from a long-term trend analysis of the entire plant to a detailed evaluation by department or by deviation category.

Summary

Accurate, consistent data in quantitative form is the basis for an auditing program. Consistency can be achieved by having one common unit of measure—deviations-from-prescribed-procedure, a fixed list of deviations, and one inspector making all of the observations. Accurate observations can be built into the program by using electrical tests to determine on a pass-fail basis whether the equipment is defective. Accurate visual checks can be achieved by using one well-trained inspector equipped with an extensive procedural guide.

The next chapter will describe how to use the reports generated from the data to manage the ESD control program. The data will provide the means for identifying problems and chart trends. The reports can be used to communicate to employees and management both successes and current problem areas. The reports to management will help to build and maintain a strong commitment for resolving ESD issues and to foster continuous improvement.

Points To Remember

- If performance is measured, ESD control will improve. Report the results, and ESD control will improve rapidly.

- Constructive enforcement and good communication helps maintain commitment to the program.

- The problems pinpointed by auditing will trigger a continual improvement sequence by always selecting the biggest problem, solving it, and then moving on to the next major problem.

- The process checkers work for the manufacturing departments. They collect data to ensure that quality is built into the manufacturing process and that ESD procedures are being followed.

- Deviations-from-prescribed-procedure is the basic unit of measure that permits valid, scientific comparisons in ESD control.

- Failure analysis and yield reports are essential, but no substitute for frequent auditing reports.

- One of the best ways to collect consistent and meaningful data is with one inspector equipped with a cart of test equipment and an inspector's guide that includes an extensive checklist of all possible deviations.

- The auditing inspector works plantwide for the quality organization. Duties include collecting data on a statistical sampling basis for audit reports. In addition, the inspector trains employees and process checkers and accepts special assignments.

- User-friendly software is needed to store data and to organize the information for analyzing and reporting.

- Auditing data converted to graphics form provides a diagnostic tool for managing the resources of an ESD control program and for communicating to management.

Chapter 9

Using Auditing Results To Manage The ESD Program

As explained in Chapter 8, auditing reports provide the coordinator with essential, timely information enabling him to understand the current status of the program. Only with this knowledge are good decisions and sound management possible. There should always be answers to the following questions: What major problems might currently stand in the way of perfect compliance or zero deviations? Were past efforts to solve major problems successful? Are resources being used to their best advantage? Are additional resources or support needed? Has there been any progress over the past two months? Can management count on further progress? Is a systemic change necessary to realize further improvement? Has management been kept informed?

The answers to these and other questions are contained in each set of auditing reports. Through analysis of the reports, it is possible to identify the most serious current problem, solve it, and then use the next auditing report to verify that the problem was solved. This process is the basis for continual improvement and makes the goal of zero deviations a real possibility. To help management understand the status of the program on an ongoing basis, send them selective reports along with a cover letter describing the efforts to reach zero deviations. ***Reporting the results in this way builds management commitment and sustains the success of the program.***

Trend and Pareto Analysis Charts

The reports are published in graphic form using the basic statistical techniques of trend and Pareto analysis charts. The trend charts reveal trends in the program by plotting deviation information for each audit cycle versus time. These charts can be prepared for the entire plant or for a single department. Figure 9-1 is a trend chart showing the trend for the ESD program over a four-year period. Figure 9-12 shows an example of a trend chart for a single department that was used to recognize and reward the department for achieving zero deviations.

The second analysis tool, equally simple and effective, is the Pareto chart. These charts are bar graphs having their data organized on the basis of an idea called the Pareto Law or The Principle of the Critical Few. Sometimes this idea is referred to simply as the 80-20 Rule. This principle, named after the nineteenth-century Italian economist and sociologist Vilfredo Pareto, assumes that 80 percent of the value of a group of items is concentrated in 20 percent of the items. According to this rule, one finds that 80 percent of the errors made by a company will be caused by 20 percent of the employees. An appropriate example would be that 80 percent of the deviations in an ESD control program will be caused by 20 percent of either the employees or the equipment.

According to this principle, one can allocate resources more effectively by first identifying what critical few efforts produce the most results and then by concentrating on those efforts. For example, focusing on the sales efforts that produce 80 percent of the profits would cut down on wasted effort. In terms of ESD control, by first identifying what critical few departments or equipment are causing 80 percent of the problems, efforts can be focused to work on solving the most serious problem in the program at the present time.

In this chapter, there are two case studies as well as additional illustrations of how Pareto analyses can be used to achieve zero deviations and to report the results to management. Case Studies 2 and 3 are good examples of how to use Pareto analyses to solve a problem.

In summary, the 80-20 Rule when organized into a Pareto chart is a very useful tool for identifying what equipment or which employees cause most of the deviations. Knowing which 20 percent of the equipment or which 20 percent of the employees cause 80 percent of the current problems permits the coordinator to focus limited resources effectively to solve the most significant problem.

When and How to Report the Results

Results ***must*** be reported to management to achieve successful communication, commitment, and continuous improvement. The importance of this cannot be overstated. Measuring the compliance with procedures will improve ESD control; however, it is the reporting of results that produces rapid improvement. The following section describes a number of ways to report results effectively.

Preliminary Auditing Reports

Within 24 hours after the inspector completes the audit of a department, a preliminary report should be sent to the shop supervisor and all interested parties for review. The report does not include graphics at this point, but rather a list of all deviations noted along with the inspector's comments and recommendations. Also included are relevant paragraphs from the Inspector's Guide describing how the tests were performed.

After the shop supervisor reviews the information in the preliminary report, he/she can speak to the ESD coordinator, not the inspector, to discuss or even challenge any item in the report. This procedure is the primary purpose of the preliminary report and is consistent with the constructive philosophy underlying the auditing program. Its intent is not to be punitive or unfair, but to work with manufacturing to identify and solve problems.

Final Auditing Reports

Those who receive the preliminary report should also receive the final report at the end of the inspection period. The report should include any adjustments resulting from discussions regarding the preliminary report. The same information printed in the preliminary report can now be finalized and put into graphic form, along with the inspection results for the entire plant, for comparison purposes.

Staff Summary Reports

Pertinent trend and Pareto charts can be selected and distributed along with a brief cover letter to provide a status report to all upper levels of management. This concise format is an effective means of keeping management informed and fostering continuous improvement.

Plantwide trend and Pareto analysis charts should always be included in these reports. For example, if major corrective action was being launched to solve a wrist strap problem, a Pareto chart would be included to highlight the problem and to solicit support in trying to solve it.

View-Graph Presentations

At the monthly manager's meeting, five minutes should be allotted to the Quality manager to report on the status of the ESD control program. Three to five charts that describe the current status can be presented in this time frame. The thrust of the talk could be to launch a major problem-solving effort, to point out how some departments have too many deviations, or to show successes in the program.

Engineering Analysis

One of the primary benefits of the graphs is that they provide an efficient means of conducting an in-depth engineering analysis. As a result, the ESD coordinator can prioritize his time and direct corrective action resources where they will be the most effective.

The graphs are also effective when meeting with engineers or shop supervisors to either review a problem or to lead the discussion in search of its solution. In addition, the graphs can provide support data for special reports to engineering or shop supervisors.

Case Study 1: Four Years Of Continual Improvement: When the first audit was completed in April 1985, the ESD control program was in place. There was a full-time coordinator, a handbook of documented procedures, all of the necessary equipment, an organized training program, and trained personnel.

Between the first and second audits, the latter being completed in July 1985, the percent of possible deviations fell precipitously from 4.3 percent to 2.5 percent (Figure 9-1). This dramatic improvement was caused, in part, by identifying and solving a serious problem with wrist straps, enhancing communication through the auditing program, and simply by having an auditing program.

How the wrist strap problem was identified and what steps were taken to resolve it will be explained in detail in Case Study 2. Solving the wrist strap problem was a major contributor to the steep decline in deviations between the first and second audit. The first audit revealed that wrist straps accounted for 37 percent of all deviations. By the time of the second audit, wrist strap deviations had fallen to 10 percent, and by February 1989, wrist straps accounted for only 1 percent of all deviations.

While the audit was successful in identifying wrist straps as a major problem, it also caused a number of other things to happen. In addition to uncovering many other problems, it also resulted in literally hundreds of questions being raised. The people in manufacturing wanted more information about ESD control. Before the first audit, it was often necessary to go into the shops and solicit questions; following it, there was rarely enough time to answer them all. Process checkers, the auditing inspector, and the ESD Coordinator became important sources of needed information.

This need for information spawned numerous announcements on the public address system and resulted in the publication of an ESD bulletin. The bulletin answered the most frequently asked questions and discussed developments in the program. Because employees had a greater need to know, training became more meaningful and effective. Supervisors also wanted to learn more. They would ask questions, get answers, and then ask more questions. The resulting dialogue encouraged them to embark on a program of self-initiated corrective actions. These new work habits of learning how to do a better job and then doing an even better job are a significant result of auditing.

The major reason for the precipitous drop in the number of deviations and for the continued decline over the next four years was due to the auditing process. The audits triggered a problem-solving, continual improvement cycle which resulted in many engineering solutions to ESD-related issues.

Except for a period of time in early 1987 when the company experienced a work-force adjustment, continual improvement has been the norm. Once the program was working, it was allowed to function and improve. It was clear by late 1987, however, that the ESD control program needed another boost similar to the one it received in April 1985 when auditing began. Case Study 3 shows how bench tops were identified as a major problem that needed to be solved and what steps were taken. Solving the bench top problem caused another steep decline in deviations, with the final point on the chart (0.3 percent possible deviations) being close to the goal of zero.

In summary, a trend chart, such as the one shown in Figure 9-1, provides the coordinator with information about trends and serves to inform management (in a succinct but powerful way) as to the history and current status of the ESD control program. Try to include this type of chart in all major reports.

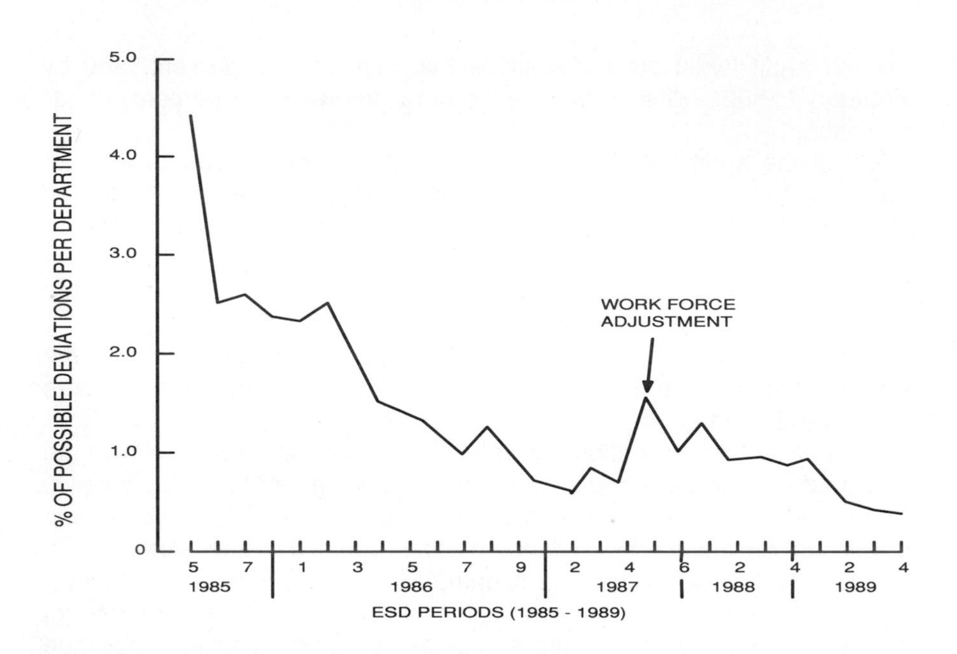

Figure 9-1. This trend chart describes—on one page—four years of continual improvement for the entire plant.

Case Study 2: Solving a Wrist Strap Problem: As stated in Case Study 1, the first audit (completed in April 1985) uncovered 4.3 percent of possible deviations on a plantwide basis. Then a Pareto analysis (Figure 9-2) revealed that wrist straps and bench tops accounted for almost 60 percent of the deviations, with wrist straps alone accounting for 37 percent.

With wrist straps highlighted as the problem most needing to be solved, a second-level Pareto chart was run as a first step in the analysis. This printout offered a closer look at how wrist straps were failing. The results, which can be seen graphically in Figure 9-3, revealed that 95 percent of the wrist strap deviations were caused by the following: open cord resistance, 35 percent; intermittent cord resistance, 30 percent; broken banana plugs, 20 percent; and loose wrist bands, 10 percent.

Further analysis revealed that many employees and process checkers did not test their wrist straps correctly. Thus, many defective wrist straps were being used. The test procedures were rewritten, and the inspector systematically went to each department in the plant and retrained employees. We also changed to adjustable, one-size-fits-all wrist straps and trained people to wear them properly.

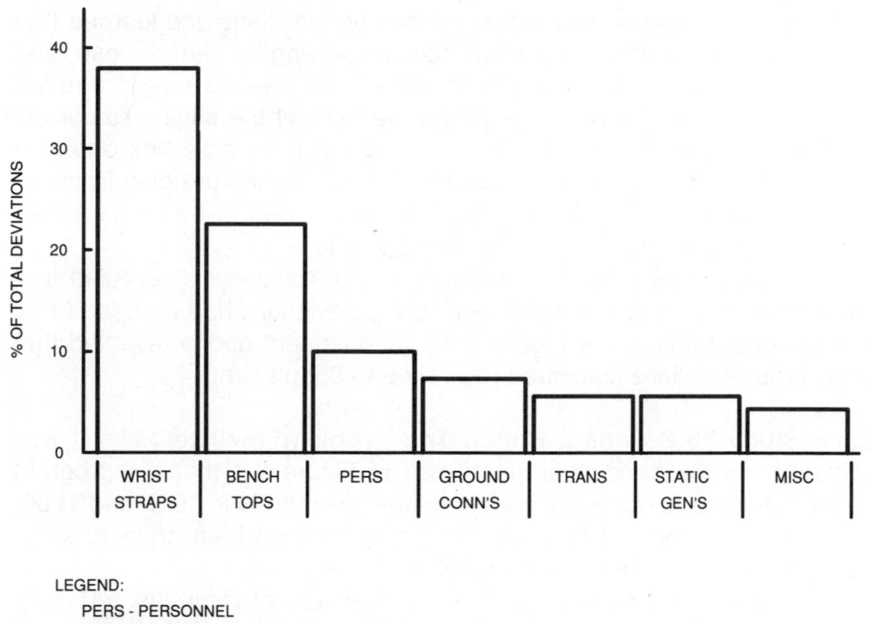

Figure 9-2. The ESD coordinator can use this type of Pareto chart to identify major problems in the program and to initiate a sequence of corrective actions. In this case, the first audit, in April 1985, revealed that wrist straps were the major problem.

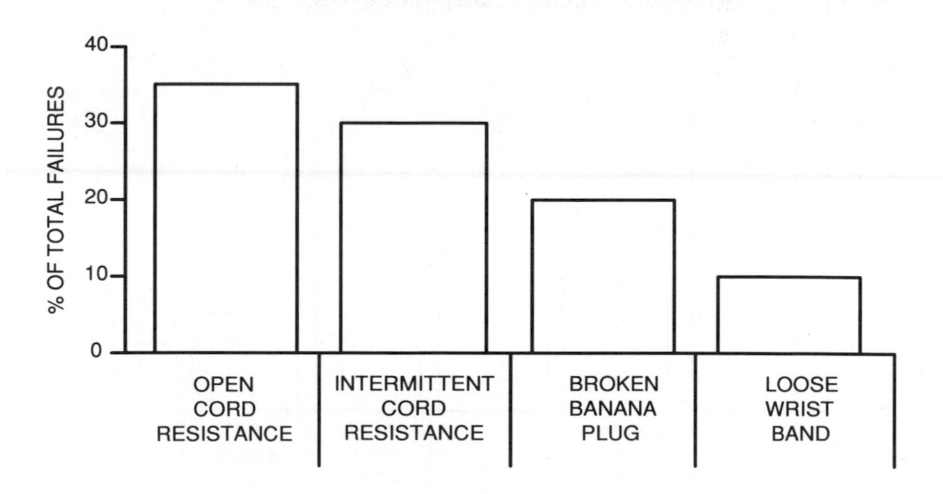

Figure 9-3. A Pareto chart of wrist strap failures

During the analysis, we studied many work positions and learned that the banana plug on the wrist strap cord was being hit, jarred loose, and damaged by the back of the bench chair. The banana plug protruded from the bench at the same height as the back of the chair. To correct this problem, we moved the wrist strap ground receptacle box one inch back from the front edge of the bench. This protected the plug from the chair. In addition, the banana plug was redesigned to be L-shaped so that it would rotate if bumped, avoiding damage.

As noted in Case Study 1, a follow-up Pareto analysis revealed that three months later, in July 1985, wrist strap deviations had dropped to 10 percent of the total. (See Figure 9-4.) Subsequent audits revealed that wrist strap deviations continued to decline to 0.2 percent.

Case Study 3: Solving a Bench Top Problem: By late 1987, it was evident from the trend analysis shown in Figure 9-1 that the program needed another major refinement similar to the efforts in 1985 and 1986. The evidence mounted that bench tops, which had been tracked since mid-1987, were the next major problem to tackle.

In addition to the standard charts, another type of comparison (Figure 9-5) was made between facility-related deviations (equipment errors) and process deviations (human errors). Human errors, usually corrected by training, accounted for over 70 percent of the deviations in April 1987.

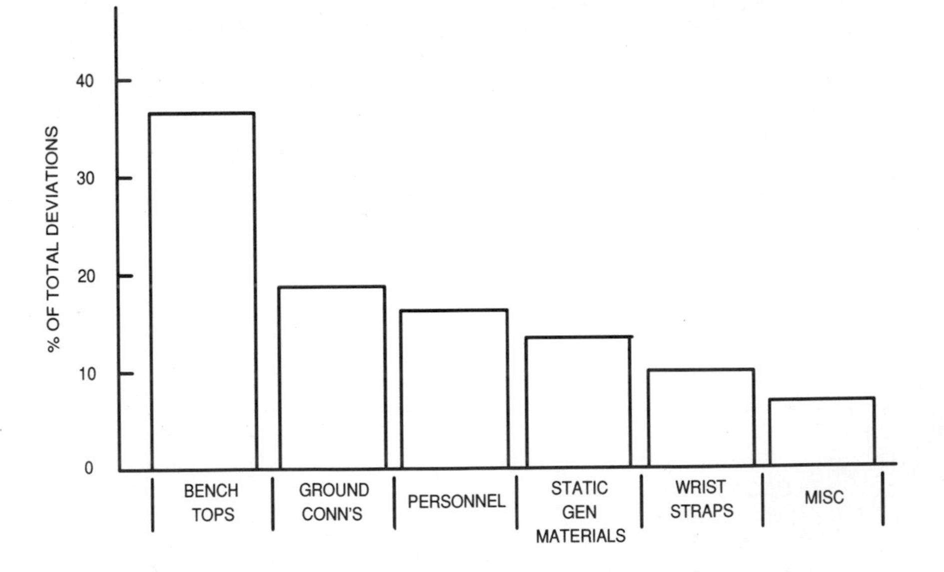

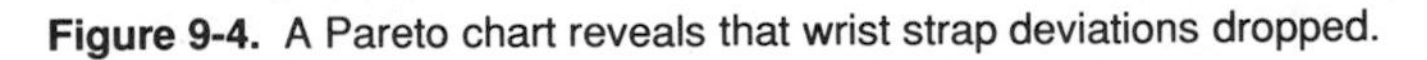
Figure 9-4. A Pareto chart reveals that wrist strap deviations dropped.

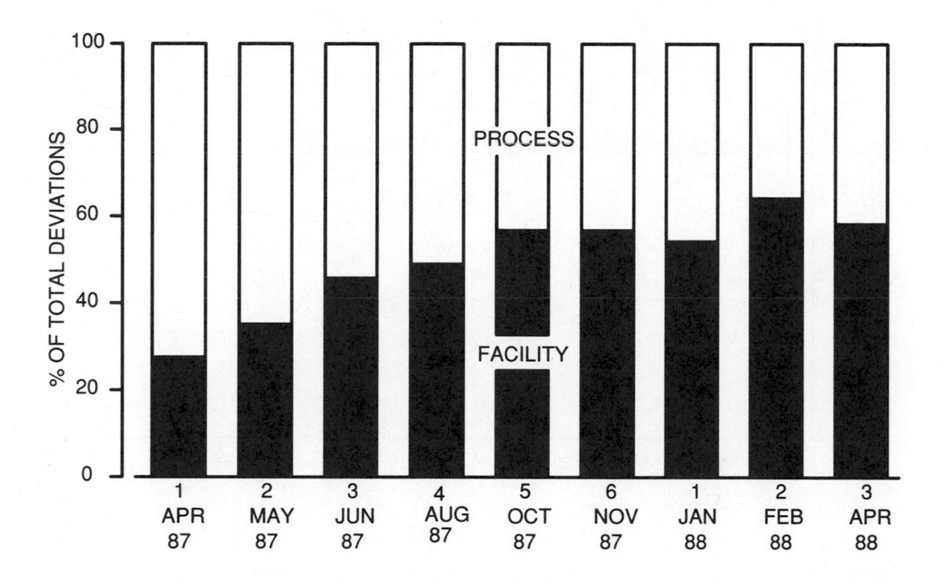

Figure 9-5. Comparing process (human error) deviations to facility (equipment error) deviations showed that, as people learned to follow procedures, equipment problems began to dominate.

As employees learned more about ESD issues, deviations from human error dropped to about 40 percent.

Thus, by February 1988, a Pareto chart (Figure 9-6) revealed that bench tops now accounted for 50 percent of the deviations. The second biggest category, static-generating materials, accounted for only 17 percent. It is interesting to note that wrist strap deviations had fallen to only 5 percent of the total at this time.

The 50 percent figure doesn't mean that the number of bench top deviations had increased. The high percentage merely reflects the fact that most of the other major problems had been solved. As a number of deviations were eliminated, the remaining ones become highlighted.

With the decision having been made to study the bench top problem further, a second-level Pareto chart (Figure 9-7) was run to start the analysis. The printout revealed that most bench top deviations were caused by the following: no ground cord, 22 percent; wrong ground bolt, 22 percent; no ground bolt, 16 percent; and shorted out resistors, 11 percent.

When we looked at the process checkers' reports, we found that (in many cases) these deviations were not noted. Tests performed by the process checkers should have caught these problems. To correct this situation, the process checker's guide was rewritten with more detailed testing instructions.

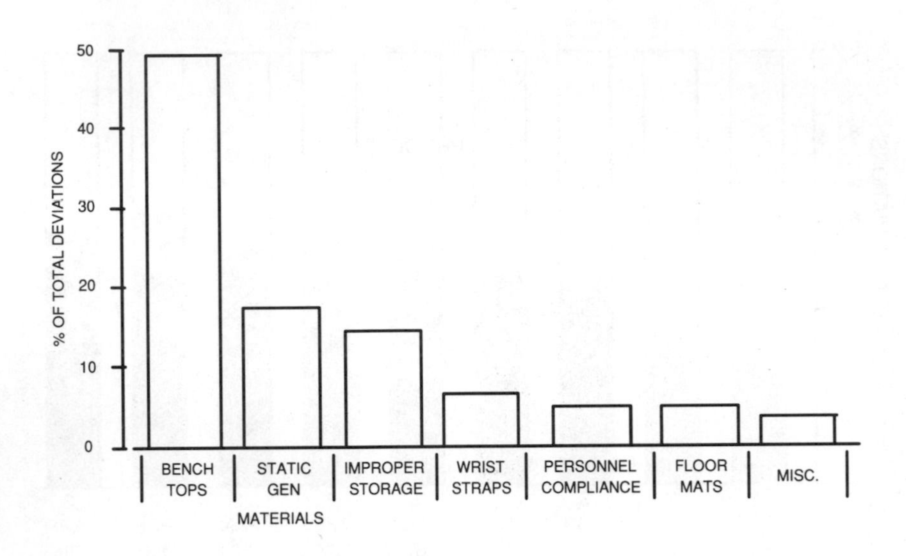

Figure 9-6. A Pareto chart shows that bench tops, which accounted for 50 percent of the deviations, were a problem that needed to be addressed.

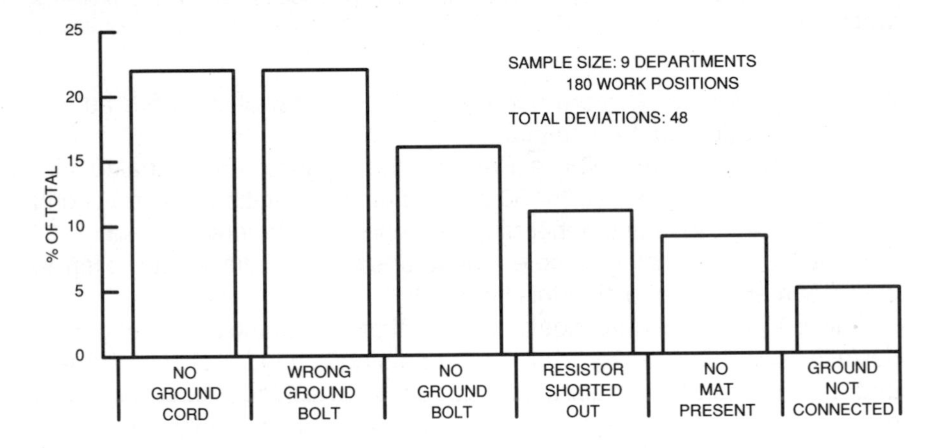

Figure 9-7. After learning that bench tops were a major problem, this second level Pareto chart serves as a diagnostic tool showing which items are causing the problem.

The auditing inspector was subsequently told to temporarily cease the audits and was placed instead on special assignment to help solve the bench top problem. One of his tasks was to visit each department and retrain the process checkers to do the electrical tests as specified in the new guide.

In addition to the process checker testing issue, the analysis uncovered a need to redesign the ground bolt and retrain the electricians on installing and testing it. We learned that the ground bolt on top of the bench was continuously being knocked off. In some cases, the bolt would not be replaced. In other cases, although the bolt was replaced, it was done incorrectly, and the electrician's test failed to detect the problem. The bolt was redesigned so that all wiring connections could be made under the bench, instead of on top. The electricians were then trained on how to install the new ground bolt and to test their work.

Subsequent audits revealed another improving trend in the number of deviations, and the Pareto chart (Figure 9-8) indicated that bench top deviations had dropped to about 3 percent of the total. The bench top problem, as with that of the wrist straps, had been solved.

In summary, the Pareto chart was found to be ***a very powerful tool*** for managing our ESD control program. It not only identified the need to initiate corrective action, it also highlighted the changes that were necessary to solve the individual problems.

Since the Pareto charts identify major problems graphically, it is simple to send this information along to management to keep them informed of the program's progress. Their understanding of the coordinator's efforts and their support in solving specific problems hasten the program's improvement. The net result is continual improvement.

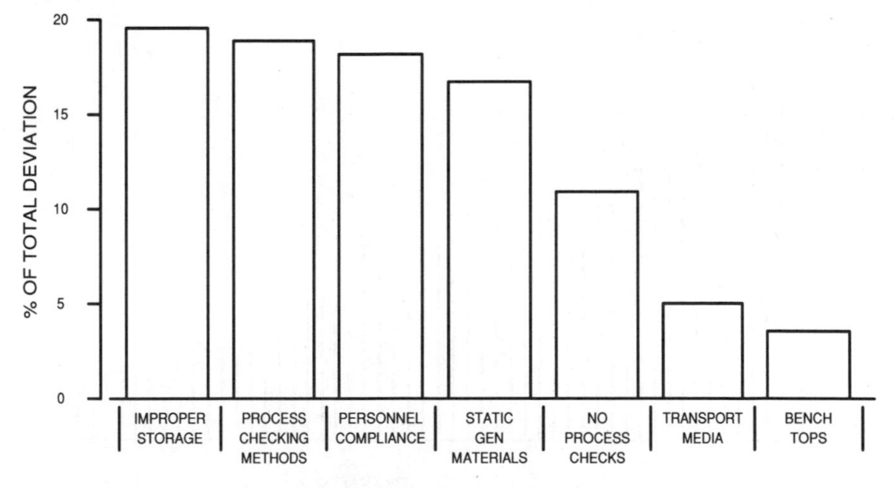

Figure 9-8. A Pareto chart printed from the next audit proves that efforts to solve the bench top problem were successful. No wrist strap failures were detected.

Illustrations on the Use of Graphics

The following illustrations provide some additional ideas on how trend and Pareto charts can be used to report results to management.

Illustration 1: Departmental Comparisons

A Pareto chart identifies which departments are causing most of the deviations. Similar to the way in which major equipment-type problems were found, one can compare all departments in the plant against the percent of the total deviations and learn where corrective action is most needed. By focusing problem-solving efforts on the approximately 20 percent of the departments who are causing almost 80 percent of the deviations, major gains can be achieved in the program. Both Figures 9-9 and 9-10 illustrate how a Pareto chart identifies the departments causing most of the problems.

By reporting the results illustrated in these charts to management, a number of different problem-solving efforts can begin. Ideally, the shop supervisor, upon seeing that his department has major problems, will initiate self-corrective action. When that happens, the ESD coordinator becomes much like a consultant working to help the supervisor achieve his goals.

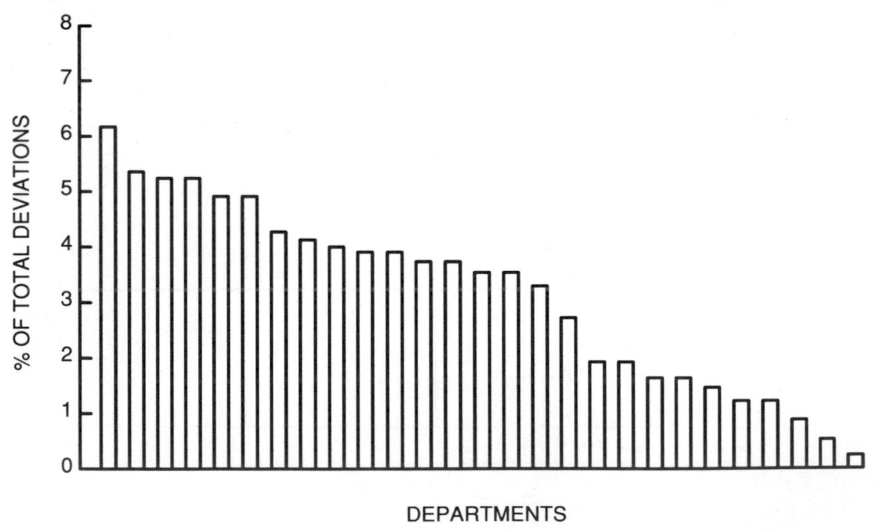

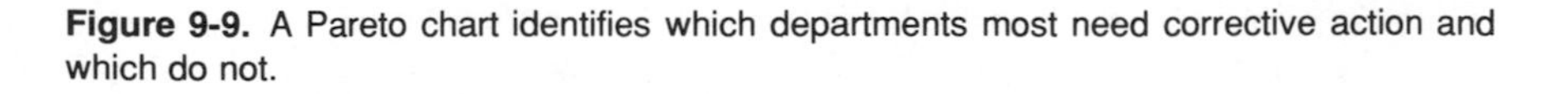

Figure 9-9. A Pareto chart identifies which departments most need corrective action and which do not.

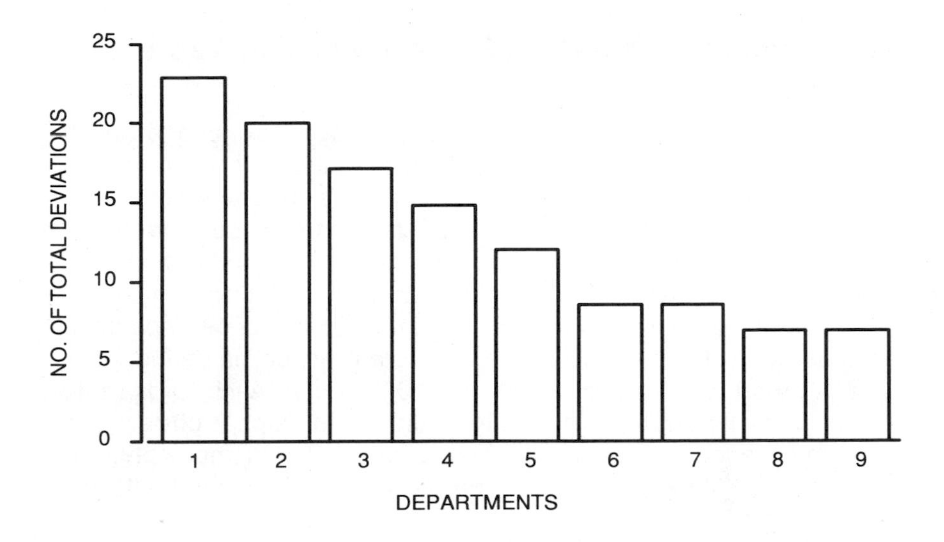

Figure 9-10. As predicted by the 80-20 Rule, these nine departments representing 20 percent of the organization accounted for about 68 percent of the deviations.

Corrective action may also be initiated by either the shop supervisor's manager or by the ESD coordinator. When the coordinator requests corrective action, a prepared form is sent to the shop supervisor, who has a week to complete the form and return it. Of course, the coordinator still serves as a consultant providing technical advice and information.

In some instances, ***manufacturing will initiate its own corrective action*** as a result of studying the Pareto chart. The shop supervisor's manager will exercise authority in an effort to achieve improved work performance.

The training manager will also study these reports and schedule retraining when necessary. It is easy for him to select the departments most in need of training because of their placement on the graph next to the Y-axis. A second-level Pareto chart will tell the training manager what type of training is most needed. In this way, scarce training resources can be used according to an intelligent priority basis. Also, the employees receive training in only those skills where improvement is most needed.

Figure 9-9 was published and sent to management in July 1985. It was part of the improvement effort that produced the first steep, negative slope in Figure 9-1. Figure 9-10 was published and sent to management in September 1988 when the program needed a push to speed up continual improvement.

Illustration 2: A One-Page Summary of Two Years Results

When managers get very busy, they tend to overlook reports and not read them. A long report is less apt to be read than a short report, and graphs can be an effective means of generating a concise report.

Figure 9-11 illustrates how ***years of data can be summarized on one page***. It contains three trend charts and three Pareto charts. The trend charts compare two years of progress by three managerial organizations, while the Pareto charts point out which of 19 departments are causing most of the problems in the current inspection period.

This report was published in mid-1987 when each organization showed some backsliding with an increase in the number of deviations. The trend needed to be reversed. By placing the Pareto charts below the trend charts, management could readily see which departments most needed to initiate corrective action.

Illustration 3: Rewarding Success With a Zero Deviations Award

Using the reports to find problems and solve them is how continual progress happens. Even though one's primary work effort is to look at problems, the progress in solving them should be recognized. It is important that supervisors and employees know when their efforts are successful. Success breeds success.

Figure 9-12 illustrates a Zero Deviations Award and a number of up-and-down periods. In the current inspection period, the department reached zero deviations and deserves recognition. When a department does achieve zero deviations, this chart is sent to the shop supervisors and is posted in the work area. It is also sent to management along with the other reports.

Illustration 4: Auditing Subcontractors or Suppliers

The work of subcontractors or suppliers can be audited using the same philosophy of auditing and reporting the results to foster improvement. In Figure 9-13, auditing to foster improvement is taken a step further.

The ESD Coordinator and plant ESD auditing inspector from our plant were assigned the responsibility of visiting the subcontractor's plant and auditing his ESD control program. In addition to the standard inspection of the manufacturing operation, we evaluated the management's commitment to ESD control. Their commitment was measured by asking fifteen questions related to the twelve critical factors described in

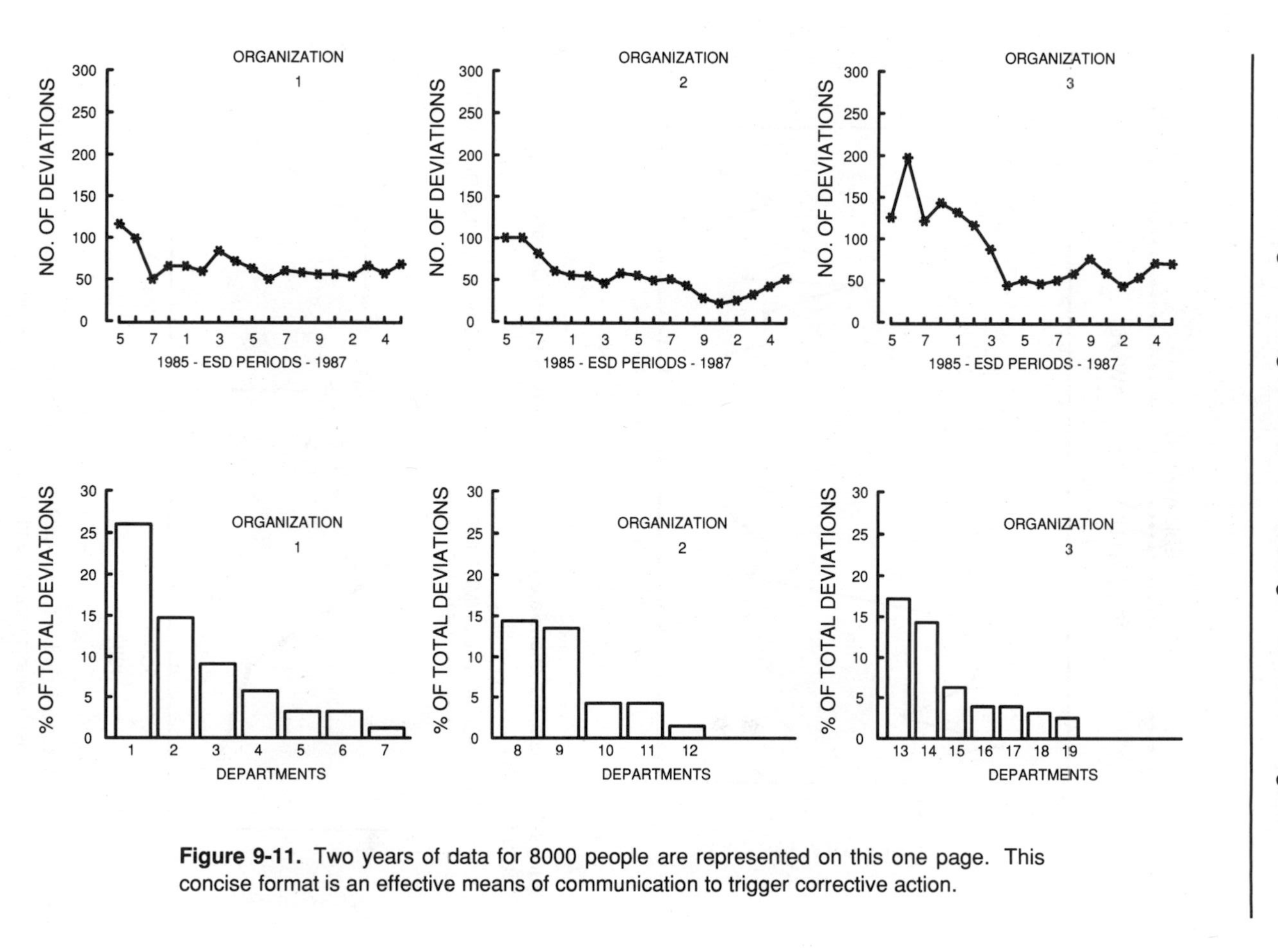

Figure 9-11. Two years of data for 8000 people are represented on this one page. This concise format is an effective means of communication to trigger corrective action.

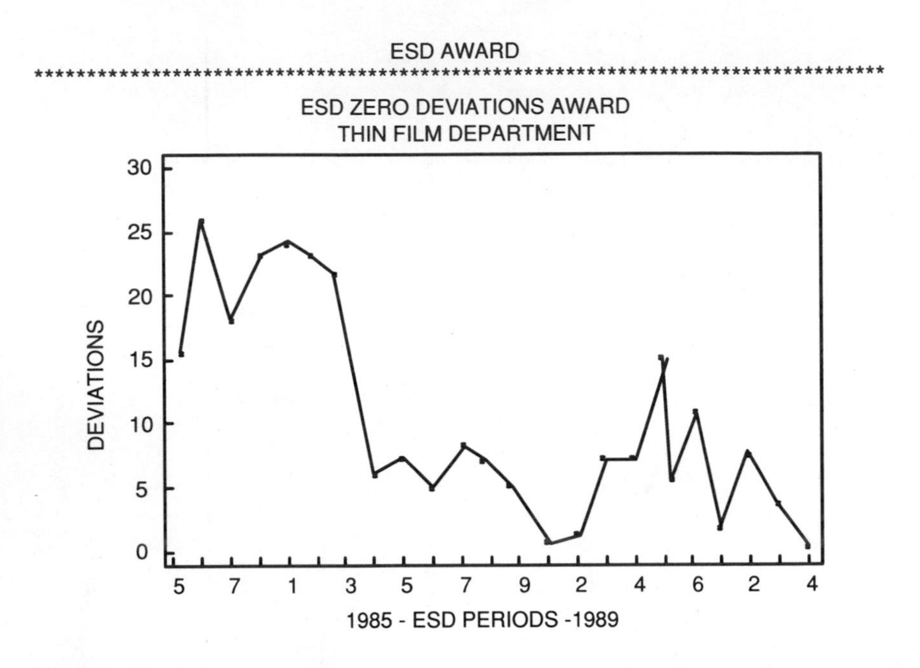

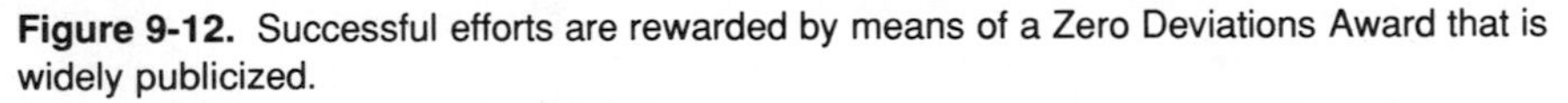

Figure 9-12. Successful efforts are rewarded by means of a Zero Deviations Award that is widely publicized.

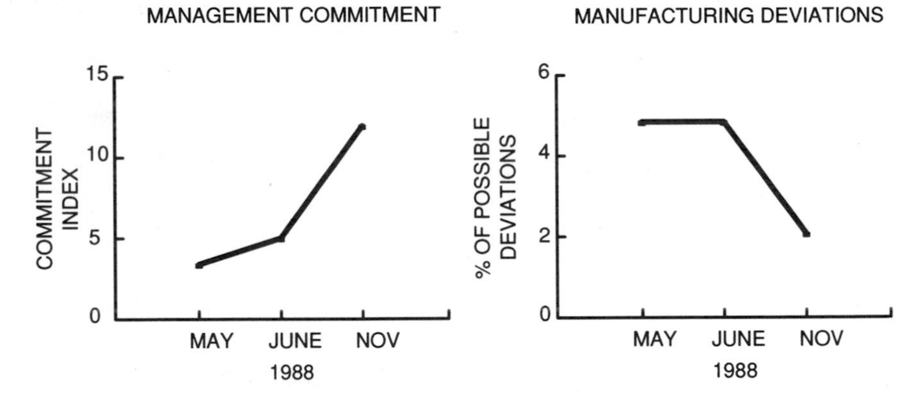

Figure 9-13. A special study of a subcontractor's efforts to improve ESD control showed that, as management commitment increased, the number of deviations decreased.

Chapter 1. These questions include: Do you have a full-time coordinator? Do you have an active committee? Do you use auditing techniques? Does auditing show an improving trend? Do you have a full-time inspector? Is training responsive to the auditing results?

The questionnaire was administered along with the first audit. They scored a three on a fifteen point scale and did equally poorly on the number of manufacturing deviations. However, six months later, management commitment measured thirteen and the number of deviations had dropped accordingly to 2 percent of the possible number of deviations.

In summary, the commitment questionnaire was an effective means of assessing the ESD control program of the subcontractor. Manufacturing deviations improved as management commitment improved.

Summary

The auditing reports provide the ESD coordinator with an important tool to manage the program effectively. Trends in the program for the plant or department can be tracked over periods of time. Major problems can be identified and priorities set. Additional graphs can be pulled from the data to begin an extensive engineering analysis. This in turns triggers a sequence of continual improvement that draws all aspects of the program into the solution of problems. In addition, any number of graphs can be assembled to give management a steady flow of succinct reports on the status of the program. As stated before, when management is well informed, they will commit themselves to the program.

Collecting auditing data and publishing the reports sets the entire ESD control program on course. Employees and supervisors will have a greater need to understand the technology of ESD control. Training becomes more meaningful. More questions are then asked and ***corrective action becomes self-initiated***.

Points To Remember

- The auditing program establishes constructive enforcement of the ESD control procedures.
- The mere presence of an auditing program motivates employees and supervisors towards better compliance. Reporting the results fosters rapid and continual improvement.
- The auditing results must be formally reported to management and employees to gain the full benefits of auditing. The reports must be concise and should consist primarily of trend charts and Pareto charts.

- The auditing reports make possible good management, diagnostic problem solving, effective communication, continuous measurable improvement, and zero deviations.
- The reports make it possible for the ESD Coordinator to efficiently do an in-depth engineering analysis.
- Trend charts reveal trends in the program. They can be used to diagnose problems and to identify when corrective action is necessary.
- Pareto charts based on the 80-20 Rule provide information for establishing priorities in the program. They also can be used to identify and permanently solve problems.
- The ESD Coordinator must initiate corrective action when the plant trend chart shows the leveling off or backsliding of deviations.
- The reports are also an effective training tool. They tell the trainers who needs training, when training is needed, and what needs to be taught. The reports can also be used to determine if the training was successful or if follow-up training is necessary.

Chapter 10

Purchasing Guidelines: Finding the Hidden Costs and Problems

The ESD control equipment (Figure 10-1) should be carefully evaluated prior to first-time purchase and, reevaluated on a periodic basis for as long as the equipment continues in service. When done correctly, the evaluation process is time-consuming and demanding. To simplify the major tasks of collecting and recording data, checklists such as those found in this chapter are recommended. In addition to the checklists, this chapter contains information and discussion on the checklist items. Thus, the pass or fail marks on the checklists should add up to defensible final decisions based on understanding and a sense of critical judgment.

In fact, a sense of critical judgment embodies the entire philosophy of purchasing presented in this book. This philosophy is summarized in the following two principles.

Purchasing Principles

Principle One: The real cost of any equipment item will differ from the label price. There are many other factors involved beyond a quoted price when tabulating the final or real cost of a piece of equipment. For example, buying the lowest priced item and then scrapping it because

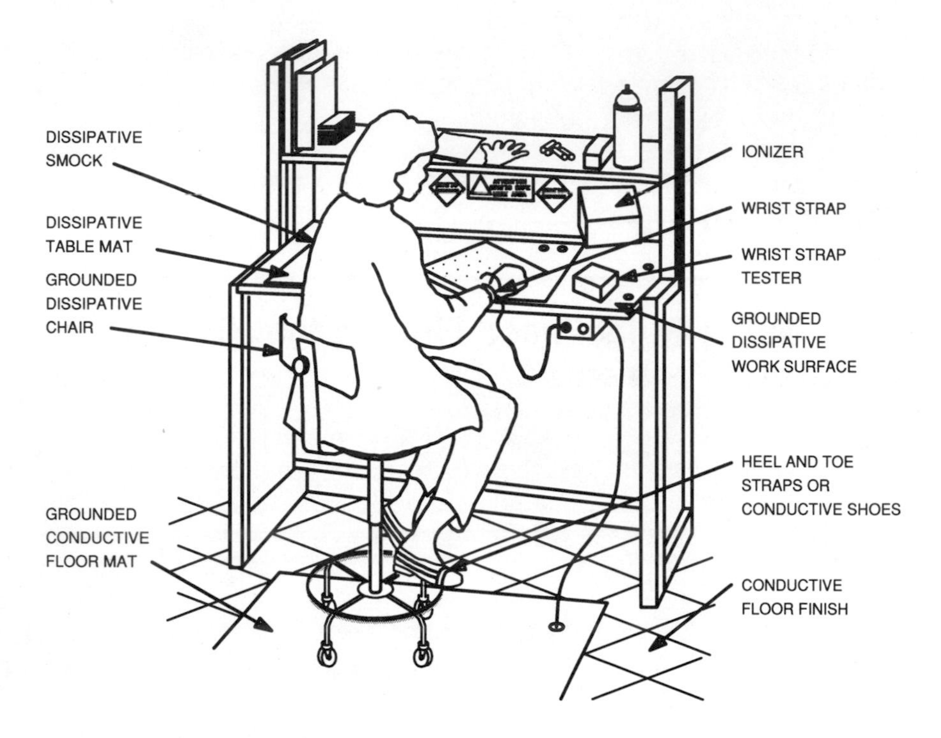

Figure 10-1. ESD control equipment at a typical workstation

nobody will or can use the item is a very expensive purchase. The final or real cost is generally more than the initial price when purchase decisions are reached superficially. Sometimes an item will not be functional when put into use. An ineffective item requires time to analyze the problem and search for a solution. This is an added cost. Then, if the item must be modified or scrapped, that is an additional cost. Therefore, employee-acceptance testing and trial testing of all items of new equipment is recommended.

The checklists will point to many other factors that can creep into either lower or raise the real cost of an item. Simple measures, such as a replaceable tab on a heelstrap or mats that can be used on either side, will extend the life of these items and consequently lower their real cost. In addition, there are some items with hidden added costs. Heelstraps that can be put on in more than one way will require additional training, which is cost, while conductive shoes require no training at all. Some table mats can be tested with a wrist strap checker while others cannot. Thus, the price of new testing equipment and training personnel to use the test equipment should be included in the real cost of an item.

Principle Two: Any piece of equipment can present a serious deficiency. Thus uncovering potential problems, most of which are hidden, is a major part of a coordinator's job. Waiting for a sudden and disastrous drop in yield is a costly and embarrassing way to learn that a piece of equipment is defective or being used incorrectly. We recommend, to avoid a problem, setting aside a fair amount of time to evaluate equipment before it is purchased. It is also recommended that the process be approached with an aggressive attitude.

The auditing reports should be studied on an ongoing basis to find hidden equipment problems. No initial screening process can be perfect, and problems will inevitably develop after the equipment has been used for some period of time. For example, an auditing report revealed that we had a serious problem with wrist straps. However, we could find no problem with the strap or the wire grounding cord. An intensive study showed that the backs of chairs were hitting and damaging the banana plug while this was plugged into the banana jack. The problem was solved by modifying the banana plugs into an L-shape and recessing the banana jack fittings under the bench top by one inch. With this hidden problem identified and solved, we removed a major cause of defective wrist straps from the program.

The process of aggressively identifying problems and solving them will force each program to change and grow in different ways. Thus the checklists presented in this chapter represent a history of our program's progress in finding and solving equipment problems. New checklist items will emerge. The lists should not be the last word on selecting equipment.

Vendor Partnership Agreements

We've learned never to buy a few wrist straps from Supplier A and the remainder from Supplier B because that type of purchasing plan allows unnecessary variations to creep into the program. Supplier A's wrist straps are inevitably slightly different from Supplier B's wrist straps. The variation, no matter how small, means that employees must know how to use two different types of wrist straps. The small difference presents an unnecessary possibility for an employee to make an unintentional error that could result in damaged devices. In addition, any slight variation in equipment forces one more detail into the training program and another item into the documentation handbook and perhaps even modifies maintenance and testing procedures.

To minimize variations in our program, we limit our equipment suppliers to ***one vendor for each type of equipment***. We have one supplier for heelstraps, one supplier for floor mats, one supplier for workbenches, and so on. In addition to minimizing variations in our program, this approach makes it possible to establish vendor partnership agreements.

With a vendor partnership, suppliers will have our business as long as they can supply the best equipment and price competitively. It is assumed that equipment changes and improvements are inevitable, and therefore, to supply the best equipment, suppliers must grow along with our program. When problems appear, they will be expected to help in finding solutions which might include adapting their equipment.

Our loyalty includes considerations such as: If another supplier approaches us claiming to have better equipment, we don't break the vendor partnership. Supplier B's equipment might be nothing more than a me-too item of equipment. We talk with our primary supplier and offer a chance to compete. We explain what improvements we are looking for and then allow some time for a researched response. If they offer an improved piece of equipment, their business success is enhanced and so is our program. In this way, the vendor partnership continues because it offers long-term mutual advantages to both of us.

When vendor partnerships are part of a purchasing plan, vendors' credentials are as important as their equipment. Thus, before any purchase decision is made, vendors should know that they will be getting a long-term customer; but at the outset, they should also know that they must meet high standards. Thus, while evaluating equipment using the criteria found in this chapter, look carefully at each vendor's qualifications (Table 10-1).

The following considerations should help in judging which supplier would best qualify for a vendor partnership.

Quality Oriented

Does the supplier ensure lot-to-lot consistency with statistical quality techniques? Does the supplier have a history of continuous equipment improvement? What initiated the improvements? What provisions does the supplier offer to guarantee continuous improvement?

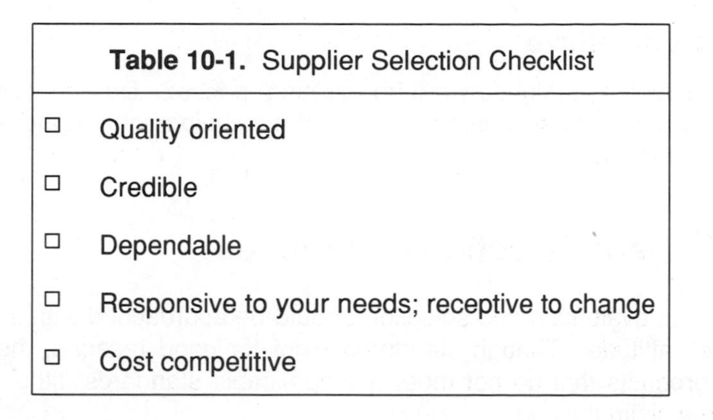

Table 10-1. Supplier Selection Checklist
☐ Quality oriented
☐ Credible
☐ Dependable
☐ Responsive to your needs; receptive to change
☐ Cost competitive

Credible

Are the representatives of the supplier honest when questioned about the technical merits of the equipment, and do they volunteer the whole story or do they conceal problems? Remember that representatives can be wrong or unaware of problems with their equipment, but still be honest. Do the representatives appear to distort the facts just to make a sale or can you rely on what they say? Do the products consistently meet supplier claims?

Dependable

Does the supplier have the ability to deliver equipment on time? Can the supplier help when special situations arise?

Responsive to Your Needs/Receptive to Change

Can the supplier offer a commitment to accommodate equipment changes as your program evolves? Has the supplier ever modified equipment based on a customer's needs? Does the supplier seem to be rigid or flexible?

In four years, due in large part to design changes, wrist strap failures at our plant have been reduced dramatically. We are now using a third generation wrist strap and have changed our supplier only once. The first wrist straps used were not comfortable and often not reliable, and sometimes body salts caused the contact resistance to increase dramatically to over 100 megohms. Our current supplier has made a number of design changes, some initiated by us and some initiated by him. Prototypes are generally delivered within a week after the changes are discussed.

Cost Competitive

Is the supplier straightforward in quoting prices? Do the supplier's estimates take into account the quality of his equipment and the realities of the marketplace?

Equipment Selection Guidelines

Equipment evaluation and selection should be approached with a "buyer beware" attitude. Though standards exist for good reasons, there are some products that do not meet the applicable standards although the suppliers claim they do.

Before you purchase any equipment, thoroughly test and qualify each item using your comprehensive guidelines and checklists (Table 10-2) created from studying this section. As part of your evaluation procedure, schedule time to observe what happens when the item is used in its intended application. After purchase, the item should be periodically retested and requalified. It should be remembered that there are no short cuts to quality control.

Table 10-2. Equipment Selection Checklist
☐ Standards
☐ Quality, Reliability and Cost
☐ Testing Compatibility
☐ Suitability for Intended Application
☐ Human Compatibility and Training Implications

Standards

What standards are available for evaluating this equipment? Will the equipment meet available standards? Appendix 2 lists the standards available to the industry. Each year the number of standards increases. However, you must still verify compliance with these standards. Don't take a data sheet at face value.

It should be noted that a complete evaluation requires a thorough understanding of these standards and how they apply to each piece of equipment. Often, portions of a number of standards are used to evaluate an item. This requires a level of expertise that a specialist can best provide. Therefore, either develop the expertise and test it in your own lab or use one of the independent labs.

Quality, Reliability, and Cost

What level of quality is needed for this equipment in this particular application? How will the long-term reliability of the equipment be determined? What will the equipment cost? The cost of an item should be considered only in terms of the quality and reliability needed. Selecting equipment based only on lower price is a false economy. You risk using less-reliable equipment that could permit ESD failures and consequently increase manufacturing costs.

When performing initial tests and periodic requalification, put quality and reliability ahead of costs. As part of your requalification program to measure long-term reliability, consider performing in-house testing and source inspection. Your ESD auditing inspector, whose function is described in Chapter 8, "Implementing an Auditing Program," can assist with periodic evaluations. Of course, if necessary, you could use independent test labs. Another source may be colleagues who have a strong quality control program and who are willing to share their data.

Testing Compatibility

Can the equipment be tested easily within the scope of your testing program? Once the item is in place, will it be tested by many people such as quality process checkers, electricians, and your ESD auditing inspector? If that is the case, you'll need to write detailed test procedures that are relatively simple and easy to communicate.

Determine whether the new equipment is compatible with your test equipment. Of course, you'll want to keep your program cost-effective by minimizing the number and types of test instruments needed. For example, try to choose control equipment, such as both a tabletop and a wrist strap, that ***can be measured with the wrist strap checkers*** that are on hand. Thus by looking carefully at the electrical properties of the control equipment, one type of test instrument can be used for a multiplicity of measurements. This will simplify training and documentation procedures.

In some instances, you will need to devise tests to be done by you, on your premises, with the equipment in actual use, to judge whether it is both suitable and compatible.

Suitability for Intended Application

Is the equipment suitable for this specific application? What is the most sensitive device that the equipment will protect? Will the equipment be compatible with other materials? For example, if you are buying a tote tray, determine whether a charge on that tote tray could be removed safely and totally when the tote is placed on the tabletop.

Figure 10-2 shows the result of a decay test used to evaluate a new tote tray with a surface resistivity of 10^{11} ohms per square. The tray was charged to 5000 volts while suspended two inches above the work surface. It was then placed on a soft dissipative (10^{9} ohms per square) table mat for approximately two seconds, and then returned to the original height. The residual voltage of 1346 volts was unacceptable. Therefore, another more conductive (10^{4} ohms per square) tote tray was selected. This time the decay test (Figure 10-3) was favorable, with a final residual voltage of 28 volts. Although the first tray met all of the applicable standards, it was incompatible with the tabletop in use. Therefore, this type of compatibility testing revealed a hidden problem that is not currently included in the industry standards.

When the equipment passes your laboratory analysis, try the item in the plant on a small scale. Quite often, subtle deficiencies will emerge as you go over the items on your list again, so it will be necessary to talk with the employees and observe how the equipment is used.

Human Compatibility and Training Implications

Will this equipment require additional employee training? Can the training be accomplished efficiently and effectively? How will the equipment affect employees? Are there any health problems associated with the materials? Will the equipment be uncomfortable or hinder employees from doing their job?

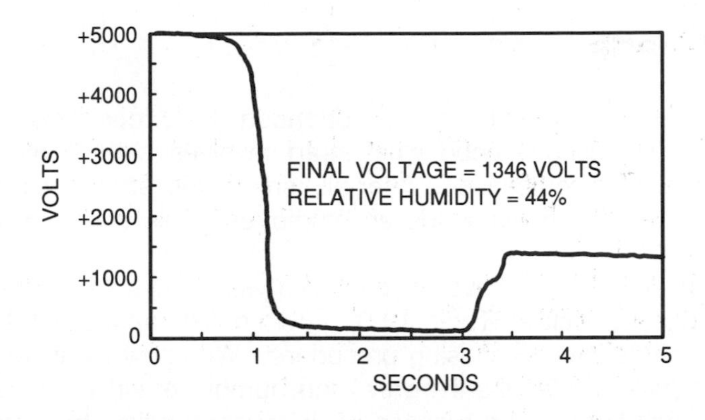

Figure 10-2. Charged plate monitor decay test of a tote tray (10^{11} ohms/square) being placed on and removed from a soft dissipative table mat (10^9 ohms/square)

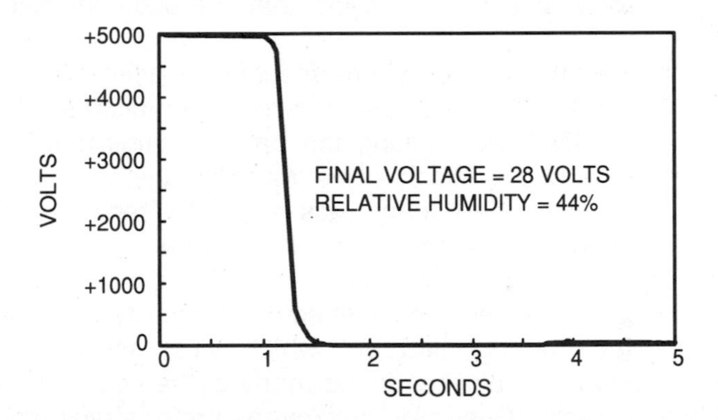

Figure 10-3. Charged plate monitor decay test of a tote tray (10^4 ohms/square) being placed on and removed from a soft dissipative table mat (10^9 ohms/square)

Will the equipment be prone to improper use? Does the design of the equipment prevent it from being worn or used correctly?

As a final check, consult colleagues in the industry. What do they recommend? Have they uncovered any pitfalls? How do they react to your major concerns? Incidentally, becoming a member of the EOS/ESD Association makes this type of communication easier by putting you in contact with many different experts in the industry. There are also a number of local EOS/ESD chapters active in the United States and Europe.

For each type of equipment, there are specific questions to be asked. Based on the guidelines above, the following examples can help you choose the correct equipment prior to purchase.

Wrist Straps

It is understandable that an inexperienced ESD coordinator might believe that selecting a good wrist strap involves nothing more than conducting a price comparison. After all, isn't a wrist strap a very simple piece of equipment that grounds an employee? This belief couldn't be further from the truth.

The problems that we have encountered with wrist straps have made the Wrist Strap Checklist (Table 10-3) grow considerably beyond a one-step cost-comparison purchasing procedure. We knew we were on the right track, because (as the list grew) the number of wrist strap failures dropped dramatically. The process of uncovering wrist strap problems and correcting them reduced wrist strap failures during the 1985 to 1989 period from 37 percent to 0.2 percent. These results let us have confidence not only in the wrist straps that we use, but also in the evaluation procedure.

Solving the banana jack problem described earlier was a major reason behind the dramatic decrease in wrist strap failures. We found that the backs of chairs were hitting the banana plugs that protruded from the front edge of the bench. This was damaging the wrist strap banana plug as well as the banana jack itself. By recessing the jack under the workbench and changing to an L-shaped banana plug, we avoided this problem. Thus, when trial-testing wrist straps, check to see whether chairs or anything else might effect the longevity of the plug.

Also during trial testing, find out which wrist straps meet with employees' approval. Employees found many of the first wrist straps to be quite uncomfortable. Generally, employees prefer straps made from fabric, but be sure that the fabric doesn't have a tendency to roll over. When a strap rolls over, the continuity is broken and, surprisingly, an employee might not notice the problem.

Employees also prefer a light coiled cord. It is possible to buy a 12-foot cord that coils to 1-1/2 feet. In addition to being more comfortable, the lighter coiled cord avoids safety problems by staying out of the way. Cords that stretch out and drape across the workbench are annoying and can present a serious safety hazard by pulling items, such as a hot solder iron, off the bench or onto the employee. This is also a good reason for under-the-bench wrist strap grounding.

A wrist strap can fail on employees who register a high contact resistance, usually due to very dry skin. Buying a limited number of metal expansion bracelets offers an effective solution. Although these wrist straps with good contact resistance are more expensive, their longevity offsets the added cost.

Table 10-3. Wrist Strap Selection Checklist	
☐	Meets EOS/ESD Standard #1-Wrist Straps
☐	Has minimum 1 megohm resistance
☐	Has a cord 6 or 12 feet in length that is lightweight and expandable
☐	Fits everybody with one expandable size
☐	Wears flat. The strap doesn't roll
☐	Has recessed rivets that are either stainless steel or nickel
☐	Disconnects at wrist
☐	Is easy to use. Clear instructions included if possible.
☐	Meets employees' acceptance
☐	Is suitably priced for quality and reliability
☐	Can be tested with available equipment and procedures
☐	Presents no training, documentation, or maintenance problems
☐	Passes trial testing program

Conductive skin creams offer another (but less attractive) solution to high contact skin resistance. The creams are an additional source of contamination and generally result in poor employee acceptance.

Select wrist straps with recessed stainless steel rivets because, after prolonged exposure, some employees will develop an allergic reaction to the nickel plated rivets. If buying wrist straps with stainless steel rivets appears impossible, at least select a wrist strap with recessed rivets to minimize skin contact.

Try to find an adjustable, one-size-fits-all strap made from a material that has a good memory. The best type of adjustable strap feature should require little effort, so that employees will not tinker with it.

Include instructions on how to make an adjustable wrist strap snug in the training program. Having adjustable straps avoids the need to stock multiple sizes and fitting employees to the correct size. Also, some nonadjustable straps stretch after use and lose good skin contact because they no longer fit.

Constant monitoring devices can be included in the purchasing selection process when dealing with certain manufacturing applications such as ultrasensitive devices or when required in a military contract. Dual or redundant wrist straps are a lower-cost alternative to constant monitoring. However, in most cases, a reliable wrist strap is sufficient protection. It should be noted that even one that fails a 10 megohm resistance test with a wrist strap checker offers considerable protection.

This can be illustrated by conducting a static decay test of a person (previously charged to 5000 volts) plugging in a defective wrist strap (Figure 10-4). A defective wrist strap can also perform surprisingly well in a float test as shown in Figure 10-5. In this test the peak voltage was less than 20 volts, whereas the same movement without a wrist strap produced 500 volts (Figure 10-6).

These results in no way preclude the need to test wrist straps frequently and to remove defective ones from use immediately. ***Wrist straps should be tested twice daily from finger tip to ground and pass a 10 megohm (maximum) upper limit***.

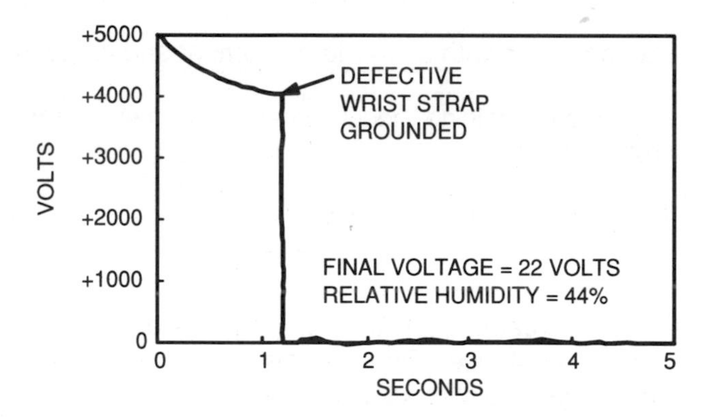

Figure 10-4. A charged plate monitor decay test of a person charged to 5000 volts and then grounded via a wrist strap that failed a 10 megohm resistance test

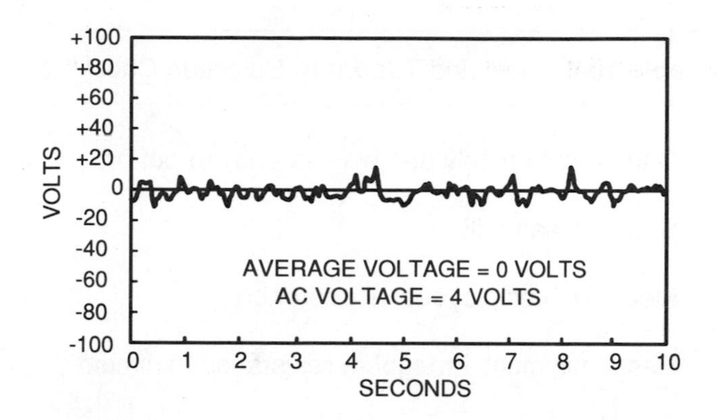

Figure 10-5. A charged plate monitor float test of a person walking on a vinyl tile floor and wearing the same defective wrist strap as in the previous figure. The peak voltage was less than 20 volts.

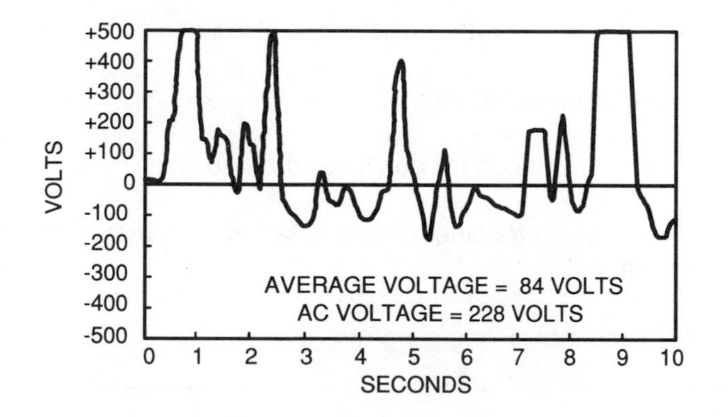

Figure 10-6. A charged plate monitor float test of a person not wearing a wrist strap and going through the same movements as in the previous figure. The peak voltage was 500 volts.

Heelstraps

Selecting a suitable heelstrap requires the same extensive effort as selecting wrist straps. The checklist has 12 items (Table 10-4).

A serious challenge when evaluating and selecting heelstraps is finding one that can be put on easily and correctly with little chance of human error. When a heelstrap can be put on improperly so that no protection is provided, a potentially serious training problem will result

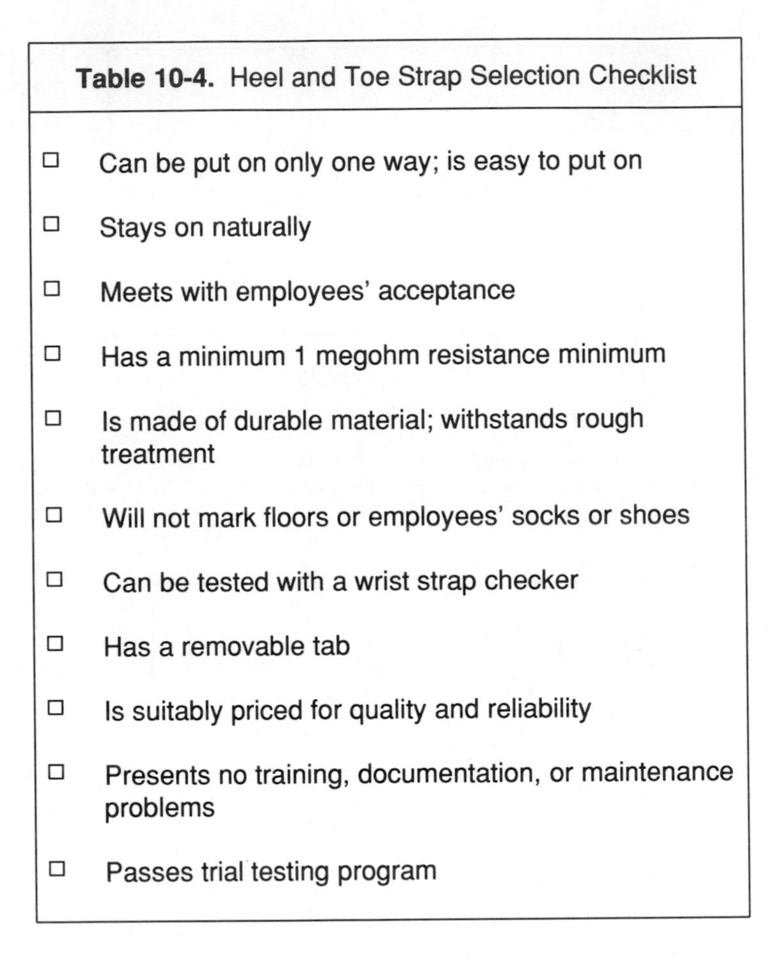

Table 10-4. Heel and Toe Strap Selection Checklist

- ☐ Can be put on only one way; is easy to put on
- ☐ Stays on naturally
- ☐ Meets with employees' acceptance
- ☐ Has a minimum 1 megohm resistance minimum
- ☐ Is made of durable material; withstands rough treatment
- ☐ Will not mark floors or employees' socks or shoes
- ☐ Can be tested with a wrist strap checker
- ☐ Has a removable tab
- ☐ Is suitably priced for quality and reliability
- ☐ Presents no training, documentation, or maintenance problems
- ☐ Passes trial testing program

and the product will experience ESD damage. Giving the employees a heelstrap that they must struggle to put on must be avoided.

Employees will point out that wearing a heelstrap that is comfortable is more important than wearing a wrist strap that is comfortable. Where an uncomfortable wrist strap is an inconvenience, standing on an ill-fitting heelstrap will cause serious pain in the arch or the ball of the foot. Therefore, extensive trial testing for employee acceptance of the heelstraps is necessary. During the trial testing, also check to see whether the heelstrap marks the floor or the employees' shoes or socks.

The potential safety problems associated with heelstraps must be evaluated and minimized. For instance, be sure that there is adequate series resistance to protect employees from danger when walking on steel floors, expansion joints, and stairways. AT&T was one of the first companies in the industry to specify this safety feature. Also, to avoid another safety problem, check to see whether or not a person might trip on the strap.

Selecting a strap with a removable ground tab will help to prevent tripping. Occasionally, these tabs become dislodged from inside the shoe and thus create a tripping hazard. However, if the tab is removable, it will release before a fall occurs. This feature also lengthens the life of a heelstrap, since the tab usually is the part that wears out first.

Heelstraps can be tested with a wrist strap checker by placing a metal plate on the floor. The continuity should be checked between the metal plate, the heelstrap, and the operator touching a wrist strap checker connected to the metal plate. However, there are some conditions that can cause an acceptable heelstrap to appear to fail the test. For example, wet leather shoes will provide a parallel path to ground and will have sufficient conductivity to bypass the resistor. Under these conditions, an employee's resistance to ground can be less than 750 kohms, which will fail the heelstrap test. Also, if the person wearing the heelstrap touches a grounded table or is wearing a grounded wrist strap during the heelstrap test, a parallel path to ground will result and can cause the same test failure.

Shoes

Unlike the serious training issues related to putting on a heel and toe strap correctly, virtually everybody knows how to put on a pair of shoes. Once ESD qualified shoes are on somebody's feet, they are virtually guaranteed to function correctly. For this reason, ***shoes are preferable to heel and toe straps***. Since carefully chosen shoes will outlast a heelstrap by a factor of three, no savings are gained by buying heel and toe straps. The checklist is contained in Table 10-5.

Shoes must fit and be comfortable and stylish to meet with employee acceptance. Therefore, extensive trial testing will be necessary. During the trial testing, the somewhat unsolvable issue of style will emerge. However, settling on one style of shoe is essential to supervision and to the ESD inspector. This makes it possible to easily identify who is properly grounded and who is not. Also, finding one type of shoe that is comfortable for everybody is unlikely, so be prepared to have some employees still wearing heelstraps, even if shoes are chosen as a primary means of grounding personnel. Without compromising comfort,

Table 10-5. Shoe Selection Checklist

- ☐ Is comfortable and safe; meets with employees' acceptance
- ☐ Durable
- ☐ Does not mark the floor
- ☐ Has 1-megohm resistance minimum and 100-megohm maximum
- ☐ Can be tested with a wrist strap checker
- ☐ Passes decay test
- ☐ Is suitably priced for quality and reliability
- ☐ Presents no training, documentation, or maintenance problems
- ☐ Passes trial testing program

be sure that the shoes are sturdy and durable. The soles should not mark the floor.

Shoes, like heel and toe straps, need adequate resistance to ground to protect employees while walking across a steel floor, expansion joints, or stairways. Furthermore, shoes that can be tested with a wrist strap checker should be selected. This will avoid the added expense of purchasing a shoe tester and the associated documentation for maintenance and, at the same time, will simplify training. These are significant benefits to keep in mind, for many shoes have a resistance either too high or too low and will require a special tester.

Shoes that exhibit these characteristics and have a resistance to ground of approximately 10^7 ohms perform well on conductive floor finishes (Figure 10-7). However, as with any footwear, they must be used with conductive flooring to be effective (Figure 10-8).

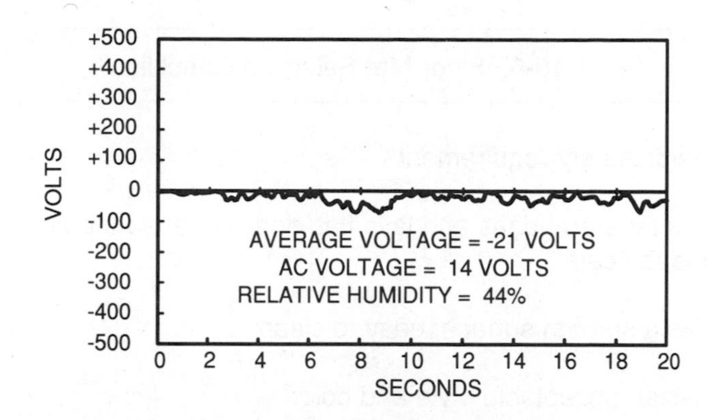

Figure 10-7. Walking on a conductive floor finish with conductive shoes generates less than 75 volts when the relative humidity is 44 percent.

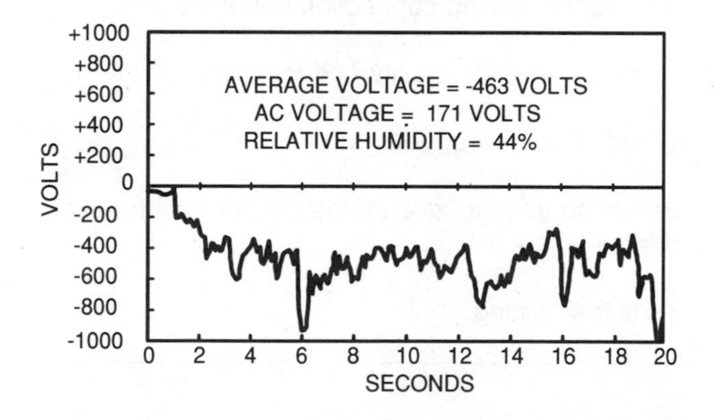

Figure 10-8. Walking on a nonconductive floor with conductive shoes does not provide adequate protection for sensitive devices or assemblies. Conductive shoes must be used with conductive flooring, floor finishes, or mats.

Floor Mats

With the expanded use of conductive floor finishes, tiles, and floors, conductive floor mats are used less and less. Although the primary purpose of a floor mat is to ground employees, in some cases a conductive floor mat is now used on a conductive floor surface to reduce fatigue. Contact the local Podiatric Society for their recommendations on fatigue mats. Table 10-6 contains a checklist.

Table 10-6. Floor Mat Selection Checklist
☐ Meets safety requirements
☐ Has tapered edges and lays flat; doesn't cause tripping or impede cart
☐ Has a smooth surface; easy to clean
☐ Has an acceptable style and color
☐ Can be tested with a wrist strap checker
☐ Has a rugged ground connection under the mat
☐ Is suitably priced for quality and reliability
☐ Will hold up under rugged use
☐ Presents no training, documentation, or maintenance problems
☐ Passes trial testing

If mats are used, choose a style that lays flat and has tapered edges so personnel won't trip and carts can move freely. The surface of the mat should be smooth. This way, dirt will not get caught in crevices and the mat can be cleaned easily with a damp mop or a broom. Of course, maintenance should trial test any floor material before a selection is made.

With people walking on mats and carts and fork trucks traveling across them, the mat material and grounding wires must be rugged and durable to ground the mats. We use braided wire similar to the type used in car batteries and anchor the wire to the floor from the underside of the mat.

A number of mats can be found with a resistance of less than 10^7 ohms. This means that they can be tested easily with a wrist strap checker because the cut-off point for a wrist strap checker is 10^8 ohms.

Workbenches

Some workbench surfaces are made from a hard laminate while others are made from a soft mat. Both types are useful. Materials will slide more easily across a hard surface, and an employee can write easily on a hard surface. On the other hand, the soft surface holds an object better and prevents slipping and sliding, and will not abrade or scratch product intended for external customers. Whether the material is hard or soft, it must be durable. Replacing bench top material is very expensive and requires extensive labor. The checklist is contained in Table 10-7.

When testing the laminated bench top, examine both the top layer and the buried, inner layer. The top layer functions as the actual work surface so it should be durable. It should be conductive enough to remove a charge from a tote tray but not so conductive as to damage a charged device by means of a too-rapid discharge.

Table 10-7. Workbench Selection Checklist	
☐	Has two groundable points; can be tested with a wrist strap checker
☐	Has wrist strap receptacles under the bench
☐	Has two or more receptacles
☐	Has a common ground point or terminal strip
☐	Is compatible with tote trays
☐	Has an acceptable style and color
☐	Is suitably priced for quality and reliability
☐	Presents no training, documentation, or maintenance problems
☐	Passes trial testing

For the best removal of a charge, we recommend that the top layer measure in the range of 10^5 to 10^9 ohms to ground. The buried, middle layer is usually conductive (measuring between 10^3 to 10^4 ohms), and must be connected to a ground for proper performance. (For more information on how to test the integrity of the ground connection on a work surface, see Appendix 3.)

Many of the hard laminates are not as effective in removing a charge as the soft dissipative mats are. For instance, when the decay test in Figure 10-3 is repeated on a hard dissipative (10^8 ohms/square) laminate (Figure 10-9), the final voltage is considerably higher. On the soft mat, the final voltage on the conductive tote tray was 28 volts, whereas here it was 263 volts. For most applications, this difference is insignificant. However, that may not be the case for ultrasensitive devices with thresholds below 200 volts. In any event, this decay test should be done before selecting any work surface or the tote trays that will be used on it.

There should be at least two wrist strap receptacles on all workbenches. The second receptacle is for a visitor or the grounding of a work cart. Both should be positioned under the bench to eliminate the cord from the work surface, and to minimize the possibility of pulling an object, such as a hot soldering iron, onto the employee. This placement also helps to protect the wrist strap plug from being accidentally jarred loose or damaged.

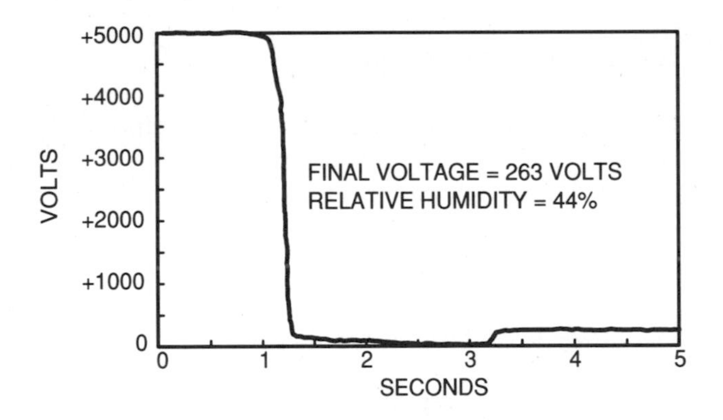

Figure 10-9. Charged plate monitor decay test of a tote tray (10^4 ohms/square) being placed on and removed from a hard dissipative laminated work surface (10^8 ohms/square).

To ensure that the workstation is at a common ground potential, be sure to specify a common ground point or terminal strip (Figure 10-10). This also provides the flexibility of adding floor mats, second shelving, or other items that need grounding. A twelve-gauge wire from the terminal to a ground is required to comply with the National Electrical Code.

The ESD coordinator should verify the safety of each application and should refer to the EOS/ESD Draft Standard 6.0, "Grounding," for recommendations on proper grounding. A resistor inserted in the path to ground of a workbench should not be considered a safety device since it could be bypassed or could introduce a safety hazard.

The workbench surface should remove a charge from the tote trays in use. Test the compatibility between tote trays and workbench surface by using a charge plate monitor to conduct a charge decay test. Another item to consider is a post-formed front edge to make the bench more comfortable. Finally, determine whether the workbench meets the cost requirements.

Work Surface Mats

The procedure for selecting mats is similar in many ways to that for selecting a laminated tabletop. Mats often have two layers, with the top layer of a mat being static dissipative and ***within the range of 10^5 to 10^9 ohms per square***. The lower limit prevents a device that has built up a charge from discharging too rapidly, while the upper limit ensures complete and safe removal of a charge in less than two seconds. The buried middle layer is typically more conductive, ***measuring between 10^3 and 10^4 ohms per square***, and needs to be effectively grounded. Table 10-8 contains the checklist.

Look for mats with a surface layer resistance to ground of 10^8 ohms so that a wrist strap checker can be used to do the testing. Even some mats with 10^9 ohms to ground specifications will pass the wrist strap test. Also, mats that remove a charge efficiently when simply placed on a grounded hard laminated surface are desirable because they eliminate the need for installing an additional ground cord to the mat.

Generally, soft mats remove charge more effectively than hard laminates (Figures 10-3 and 10-9). The compliant nature of a soft mat results in a lower contact resistance and thus better charge removal properties. However, either material is adequate for most applications. Soft mats should be considered for ultrasensitive applications where the device thresholds are below 200 volts.

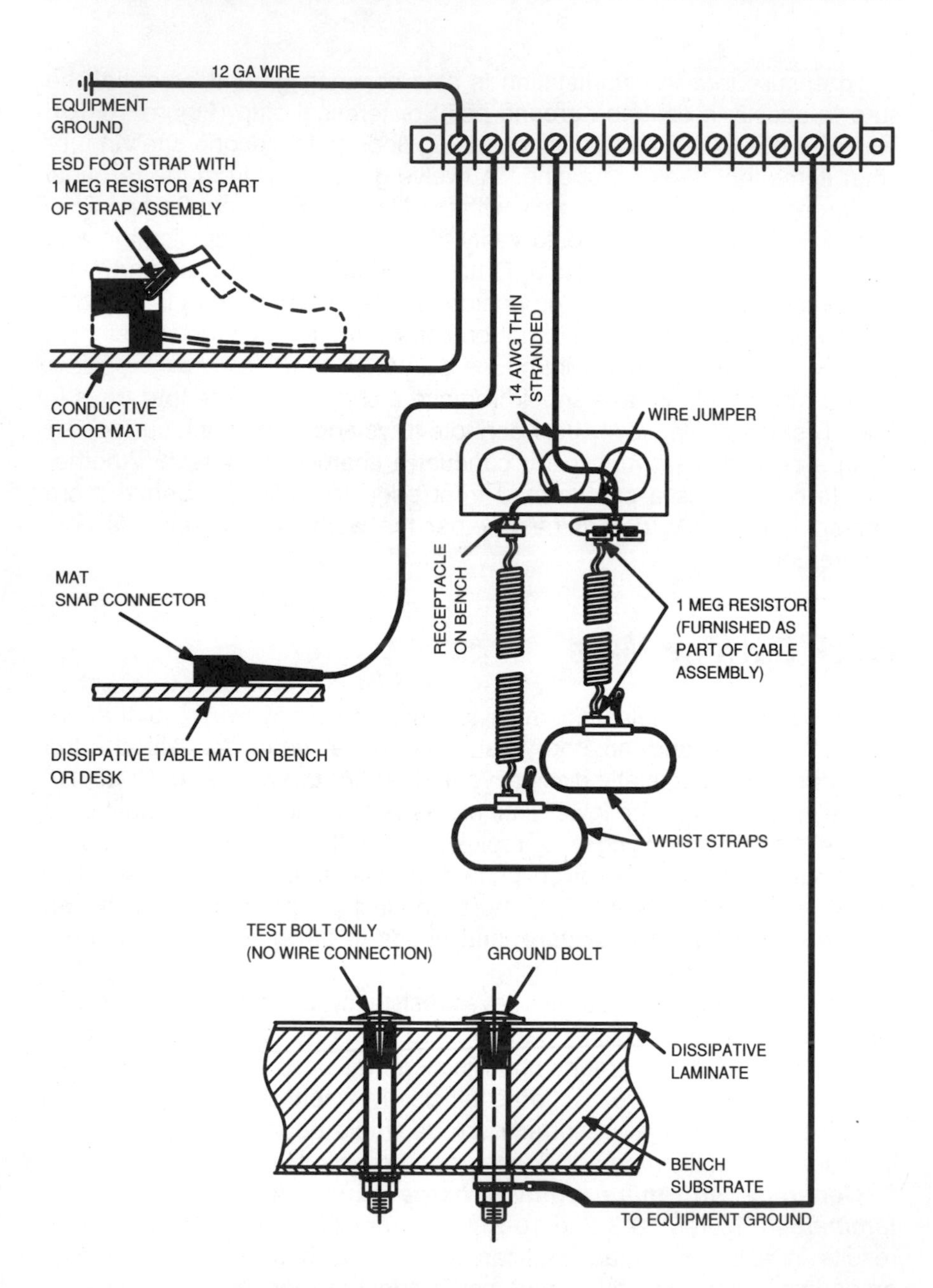

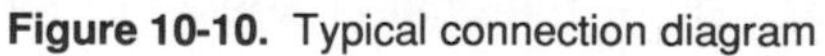

Figure 10-10. Typical connection diagram

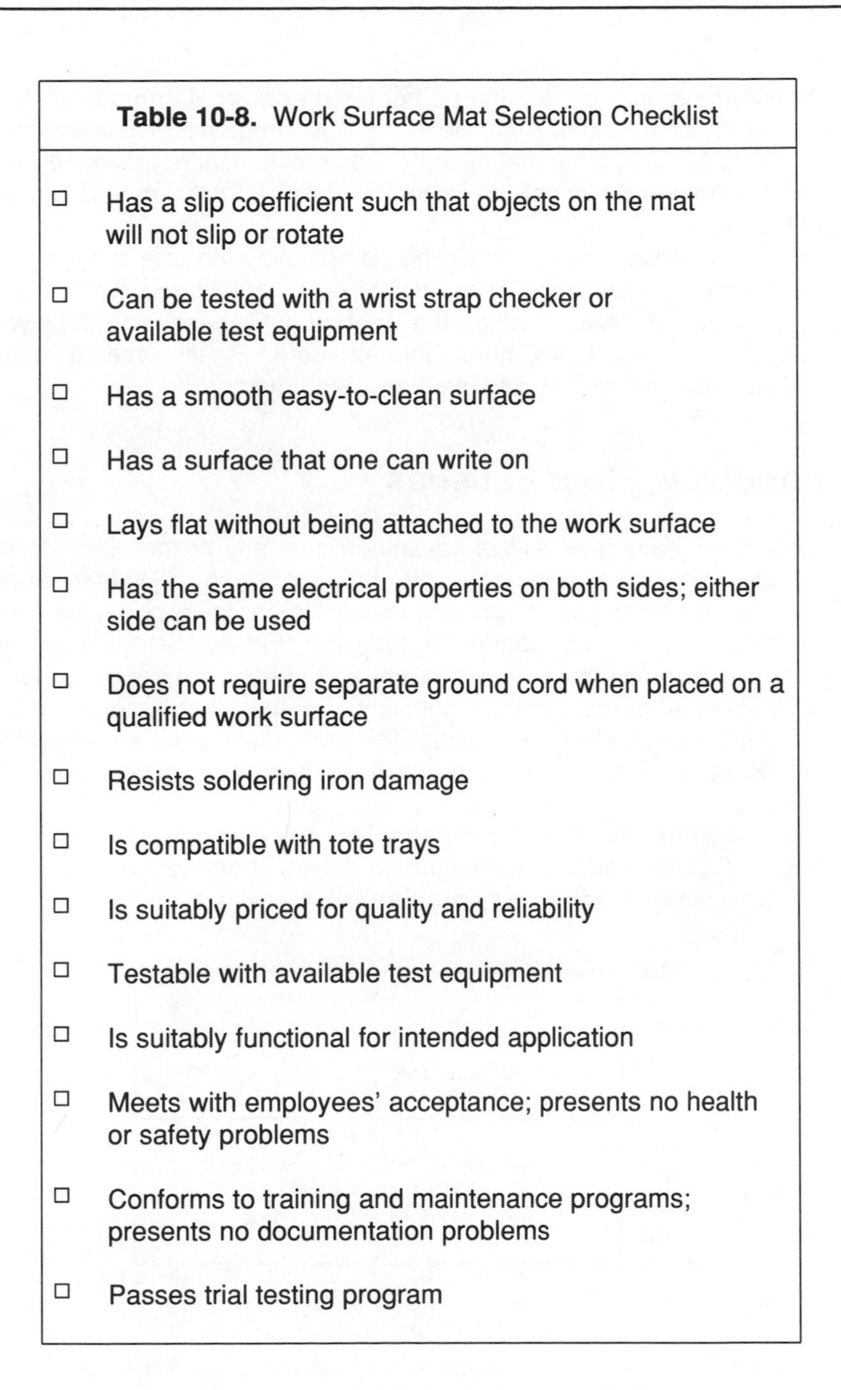

Table 10-8. Work Surface Mat Selection Checklist

- ☐ Has a slip coefficient such that objects on the mat will not slip or rotate
- ☐ Can be tested with a wrist strap checker or available test equipment
- ☐ Has a smooth easy-to-clean surface
- ☐ Has a surface that one can write on
- ☐ Lays flat without being attached to the work surface
- ☐ Has the same electrical properties on both sides; either side can be used
- ☐ Does not require separate ground cord when placed on a qualified work surface
- ☐ Resists soldering iron damage
- ☐ Is compatible with tote trays
- ☐ Is suitably priced for quality and reliability
- ☐ Testable with available test equipment
- ☐ Is suitably functional for intended application
- ☐ Meets with employees' acceptance; presents no health or safety problems
- ☐ Conforms to training and maintenance programs; presents no documentation problems
- ☐ Passes trial testing program

Specify mats that lay flat and do not need additional grounding when they are on a grounded laminate top. This simplifies installation and maintenance. Also, mats that have the same electrical properties on the top and bottom are desirable because they can be turned over and used on either side.

As with tabletops, mats should be compatible with tote trays, resist damage from hot soldering irons, and have a nontextured surface that can be cleaned easily. Also, the mat should pass an employee acceptance test and should be durable and safe. Finally, when all other criteria are met, the mat should meet cost requirements.

Conductive Floor Finishes

Floor finishes have one distinct advantage over any of the alternatives. They are ***inherently antistatic*** and, thus, forgiving of human error. Conductive floor finishes will prevent excessive charging of people even when they forget to wear conductive footwear (Figure 10-11). A person walking on a conventional floor wax on a vinyl floor will generate up to 5000 volts at 40 percent relative humidity. On the other hand, a person walking on a conductive finish under the same conditions will generate only 200 volts. This feature is a significant benefit of conductive floor finishes.

Floor finishes are also considerably less expensive than conductive floors. In fact, we saved over five million dollars at one location by using the floor finish instead of replacing existing tile with a conductive floor tile.

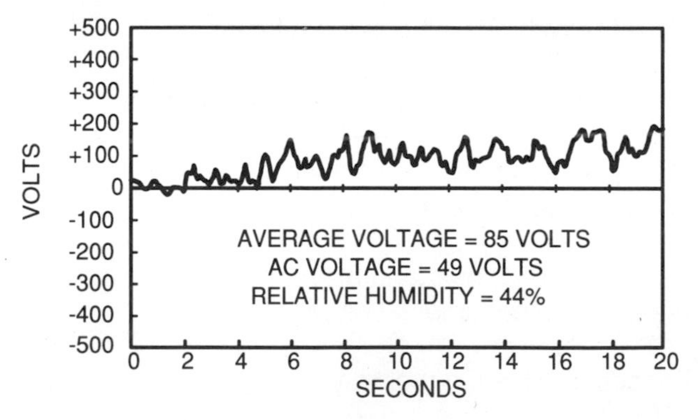

Figure 10-11. Walking on a conductive floor finish with nonconductive shoes

This savings included the material and labor that would have been associated with removing the old tile and installing new conductive tile.

However, floor ***finishes require continued maintenance which must be done correctly***. If put on too thin, the electrical properties will be compromised. Training maintenance employees in how to apply the finish and adding additional periodic testing of the finish are items that must be factored into choosing finishes over conductive floors or mats. Also, the floor finishes on the market vary in performance, so careful screening is necessary. The selection checklist is in Table 10-9.

To weed out deficient products, test the electrical properties of the surface thoroughly.

Table 10-9. Conductive Floor Finish Selection Checklist	
□	Has a high slip coefficient so that employees will not slip
□	Wears well; doesn't lose electrical properties from normal wear
□	Meets management acceptance for appearance
□	Meets maintenance department acceptance
□	Resistivity less than 10^{10} ohms/sq; has acceptable decay properties
□	Tests well for antistatic properties
□	Is compatible with footwear worn in the plant
□	Meets employees' acceptance
□	Is priced suitably for quality and reliability
□	Presents no training, documentation, or maintenance problems
□	Passes trial testing

Testing should be done before use, and trial-testing should be done after people have used the surface—this means that the surface should be tested after it has been walked on, scratched, dirtied, damp mopped, and maintained. To remove a charge, the surface should have a surface resistivity of less than 10^{10} ohms per square. Test the surface in a number of places to see whether there are any blank spots. Also, the surface should resist tribocharging. Have personnel wearing the type of shoes used in the plant walk across the floor and, using a charge plate monitor in a floating mode, test for static charging.

In the employee acceptance test, include a check to be sure that the floor is not too slippery. Improper application of a floor finish can make some acceptable floor finishes excessively slippery.

Include maintenance people in the selection process of a walking surface. The ideal finish should be easy to apply, should maintain electrical properties, and should look attractive after buffing. Ask Maintenance to see whether a finish is difficult to apply or maintain.

Does the finish have a high drag coefficient when applied with a mop? Does it bubble? Can it be damaged by water? Is the finish as attractive as conventional surfaces? Will the electrical performance hold up over time? Of course, cost is a final consideration in the selection process.

Bench and Room Ionizers

Generally, ionizers are not needed in a comprehensive ESD control program since passive techniques will solve most problems. Avoiding the purchase of an unnecessary ionizer will save money as well as eliminate another item to maintain. This is an opportunity to keep your program realistic. However, there are some instances where an ionizer is the only solution to a problem. A usable rule of thumb is that approximately 5 percent (or less) of the workstations will need an ionizer. Table 10-10 contains the checklist for ionizers.

Ionizers should be used in clean rooms, in conditions where wrist straps are not viable and static-generating containers are necessary for handling sensitive devices. Ionizers are useful where circuit boards are taped or where static-generating material must be combined with the boards. When handling ultrasensitive devices, balanced ionizers provide the additional protection that is often required.

However, it should be noted that an ionizer can actually damage an ultrasensitive device rather than provide protection. This might happen if the offset voltage is higher than the threshold of a device. For example, a device with a 20-volt threshold can be damaged by an ionizer with a

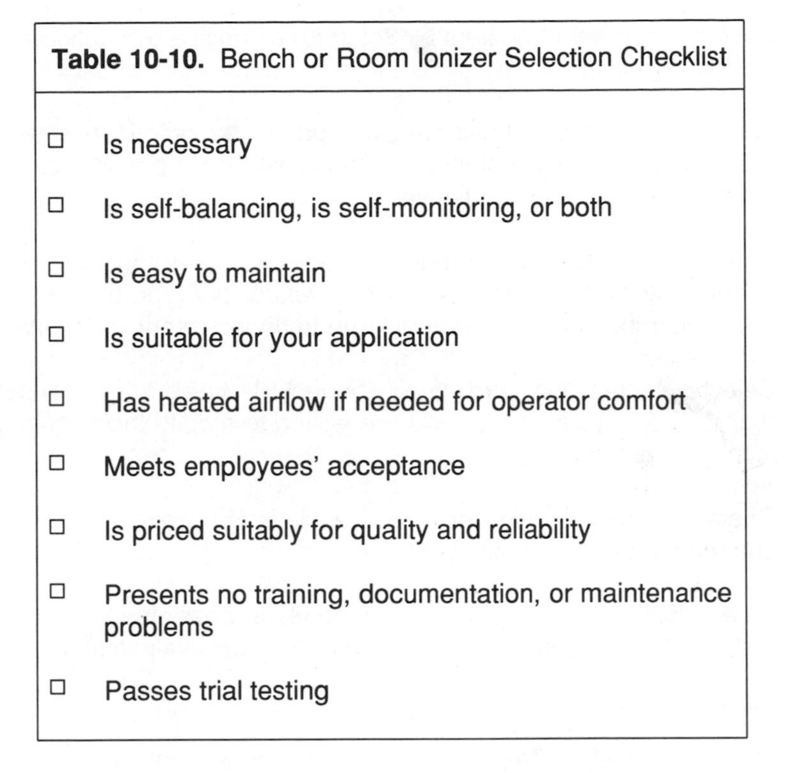

Table 10-10. Bench or Room Ionizer Selection Checklist

- ☐ Is necessary
- ☐ Is self-balancing, is self-monitoring, or both
- ☐ Is easy to maintain
- ☐ Is suitable for your application
- ☐ Has heated airflow if needed for operator comfort
- ☐ Meets employees' acceptance
- ☐ Is priced suitably for quality and reliability
- ☐ Presents no training, documentation, or maintenance problems
- ☐ Passes trial testing

70-volt offset voltage. Offset voltages of this magnitude are more likely to occur when using a room ionizer in a room without laminar flow because of the uneven ion dispersement created by the irregular airflow. An out-of-balance ionizer can damage a sensitive device in the same manner.

Packaging and Handling Materials

See Chapter 12 for an in-depth discussion on packaging considerations.

Points To Remember

- Selection of ESD control equipment has a very high priority.
- Develop checklists and written guidelines that approach equipment selection as a science.

- Find the real cost of an item by doing an extensive evaluation to find the hidden costs.

- Find the hidden problems with equipment by doing an extensive evaluation prior to a first-time purchase and on a periodic basis for as long as the equipment is used.

- Establish long-term agreements, called vendor partnerships, with vendors by limiting purchases to one vendor per type of equipment. This will reduce training problems due to unnecessary variations.

- Select vendors that are quality oriented, credible, dependable, responsive to your needs, and are willing to modify their equipment to solve a problem.

- Change vendors only for an overwhelming advantage. Never change for a me-too equipment decision.

- Select ESD control equipment, whenever possible, that can be tested with a wrist strap checker during periodic evaluations.

- Always consider the two purchasing principles:

  Principle one: The real cost of any equipment item will differ from the label price.

  Principle two: Any piece of equipment can present a serious deficiency. Therefore, a thorough evaluation is essential.

- If necessary, devise a compatibility test. (It may not be included in the industry standards.)

Chapter 11

Training for Measurable Goals

Does training make a difference? It certainly does! Our experience shows that employees who are trained to comply with ESD control procedures do a better job. The evidence comes from our auditing reports. They show that ***untrained employees account for most of the deviations***, while trained employees cause very few deviations. For example, during one auditing period, four out of every five deviations were caused by employees with little or no training (Figure 11-1). Subsequent audits revealed that when employees received training or were given additional training, the number of deviations in those work areas decreased dramatically.

Reducing the number of work performance deviations to a minimal or (even a zero) level is the ultimate goal of a successful training program. Considering the large numbers of employees to be trained, this is a formidable challenge. It requires much more than supplying equipment and issuing fiats such as "wear a wrist strap" or "put sensitive devices in a bag." Fortunately, the challenge is made somewhat easier because most employees are more than willing to comply with procedures when offered the right training at the right time. This chapter explains how to do this.

Part One of the chapter describes how to write measurable goals while preparing some of the reference materials for the program. This should be done prior to teaching any classes. Carefully study the Three

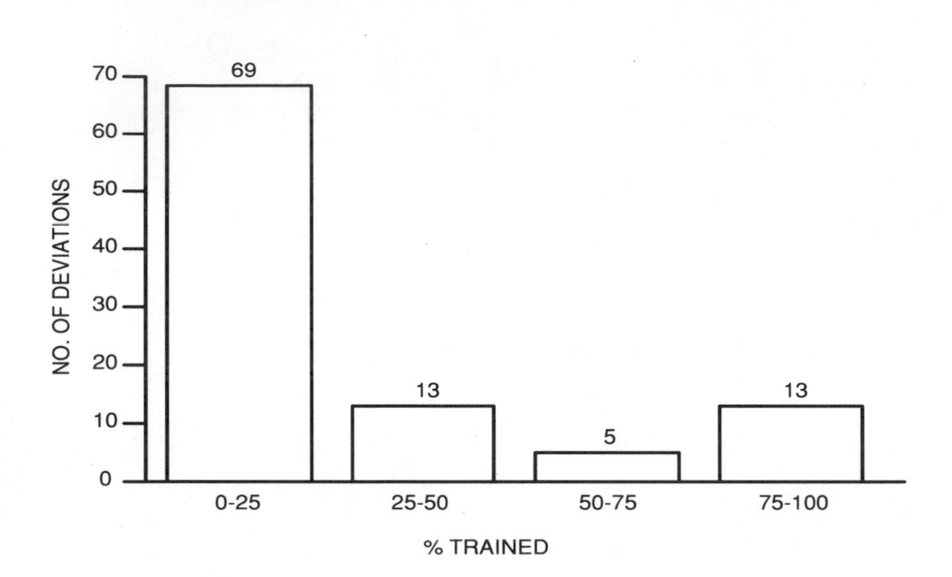

Figure 11-1. A basic frequency distribution reveals that trained employees make few errors while untrained employees account for most of the deviations from proper procedures.

Principles of the Psychology of Training and Learning[1] explained in this chapter. The Principles describe how people learn and how the program's goals can be written to both enhance learning and to measure the program's effectiveness. The first of the Three Principles states that one should train only to affect a measurable change in work behavior. Study this Principle carefully. It touches all aspects of the program.

The next step is to set out the program's goals based on the documentation guide and the course outlines. The documentation specifying work procedures, goals for each class, and auditing checklist items are woven from the same thread. Thus, when the documentation is written in work performance language, the statements can be easily translated into performance-type goals for each training class and into checklist items for auditing the training program. In addition, if the documentation is written in easy-to-understand language, it will double as a reference guide.

Part Two of this chapter describes what courses to schedule, how to prepare specific classes, and suggested training techniques. The list of recommended courses reflects the need to schedule some type of goal-oriented training in ESD control for almost every employee in the organization. At AT&T, we offer courses to eight different groups of employees from bench employees to senior-level management. We also offer Awareness Training and an annual Quality Fair for all employees.

Part One: Setting Measurable Goals

Study The Three Principles of The Psychology of Training and Learning

Principle One: Train only to affect a measurable change in work behavior. This means that the trainer will direct all instruction to some observable outcome described in the handbook which will help the employee to do his/her job better. For example, the goal of instruction will be for the employee to demonstrate that he/she can wear a wrist strap correctly, detect a worn wrist strap, or test the wrist strap. The trainer can demonstrate the skill to be learned to the class. The employee can follow the instruction in class and do follow-up practice at the workbench under the guidance of a supervisor or the auditor.

Not all goals need to be physical. For example, a sensible goal would be to have the employee explain how ESD damages devices of varying thresholds or tell what materials are most apt to carry a high level of charge. Questions can be asked and discussed in class with follow-up conversations between the employees and their supervisor or the ESD auditing inspector. The measurable change in behavior is how well the employee can answer questions and talk about ideas discussed in class.

These goals cast a different light on traditional instruction. In a traditional training session, a trainer might tell the audience that ESD damages devices and wrist straps ground personnel. A written test might be given. However, unless the instruction successfully affects a measurable change in work behavior, valuable work hours have probably been wasted. Worse than the wasted hours in training is the illusion that employees are complying with ESD control procedures.

It will be important to write and think in terms of work performance-measurable goals. Using the following goals for wrist straps as a sample, make measurable goals for the following areas: footwear, static generating materials, triboelectric charging, device failure, and transport devices.

Goals for Wrist Strap Training:

1. The employee will be able to put a wrist strap on and adjust the strap so that there is good contact between the skin's surface and the strap.
2. The employee will be able to explain how a wrist strap removes a charge.

3. The employee will be able to distinguish between a worn wrist strap and a good wrist strap.
4. The employee will be able to use a wrist strap checker to determine whether a wrist strap is working properly.

Principle Two: Motivate students to improve learning. Psychologists have proven that motivation improves learning. Students show better retention of material and remember the information and skills longer. Also, psychologists point out that motivation has many facets in addition to attention-getting techniques.

First, students tend to be better motivated when they understand the purpose behind the class. Thus, be sure to explain how ESD control helps the company to build and sell a better product. Explain how they get credit for doing a better job. It might be helpful to show that when devices are found to fail final tests without any known cause everybody is blamed. Knowing though that ESD has been eliminated as a source of damaged devices removes false finger pointing. Yield charts, showing how failure rates decline with increased ESD control, are very convincing.

Then, be sure to relate in concrete terms, how the goals of the classes will help each person to do a better job while building a better product. Help them to see that wearing a wrist strap correctly makes a difference. This can be done by demonstrating how devices fail when a person does not wear a wrist strap and how wearing the wrist strap prevents failures.

In addition to helping students understand the purpose of the classes, prepare the lessons with a specific class of students in mind. Make the material appropriate to their level. Keep the lessons concrete and familiar to the students whether they are bench employees or engineers.

Genuine compliments and the feeling of belonging that occurs when people are aware that everybody is pulling together helps to motivate students. Tell employees in class or over the speaker system when goals have been met. When the goals are concrete and measurable, compliments can be honest and sincere. Avoid negative motivation by not intimidating or humiliating employees either in the classroom or on the job. This technique fractures the group rather than builds group spirit. Don't criticize one person in front of others. Remember, there is no sense in asking a question when everybody knows they don't know the answer.

Classes where learners can participate by touching things, seeing demonstrations, or engaging in discussions help motivate learners. Since electrostatics has so many interesting and startling demonstrations, be sure to build a repertoire of demonstrations into the ESD control training program. A number of demonstrations are described in detail later in this chapter.

Principle Three: Take into consideration that students tend to forget information and skills that are not used regularly. Stated another way, we remember best that which is used immediately and then practiced with great frequency. Although there are a few exceptions to this rule which will be explained later, schedule courses only when employees need to learn some skills to be used on the job. During the class, have the students practice the skill being in the area learned. Then be sure there is immediate follow-up learning where the person works.

Derive Measurable Goals Based on the ESD Handbook

Base all goals and course development on the specified ESD Handbook that was prepared with training in mind. The AT&T "ESD Control Handbook" includes the following topics:

- Basic ESD Concepts
- ESD Failure Models
- Sources of ESD Damage
- ESD Effects (Dead-On-Arrivals, Device-Operating-Failures)
- ESD Control Techniques
- ESD Control Committees
- ESD Problem-Solving Guide
- Definition of Terms.

The Handbook should detail the procedures that must be followed. If the finished product is written in easy-to-understand language and has good graphics, it can serve not only as a handbook, but also as a training document and as an on-the-job reference manual.

Since the measurable goals and the training goals are based on the ESD Handbook, begin by reviewing it. It is also helpful to review the Auditing Checklist. The teaching goals can be derived from both of these documents.

In fact, the teaching goal is generally self-evident once the Handbook and the Auditing Checklist have been reviewed. The following is an example of how teaching goals were derived from our Handbook and Auditing Checklist.

Handbook Statement:

Wrist Straps. "The sole purpose of wrist straps is to ground personnel...Continuity from point of connection to ground must be maintained at all times."

Auditing Checklist Item:

Personnel not wearing wrist strap properly.

Personnel not grounded.

Wrist strap failed test.

Teaching Goal:

The student will know how to put on a wrist strap correctly and adjust the strap so that it fits snugly against the skin.

The student will be able to explain how personnel grounding helps prevent devices from being damaged.

The student will understand the importance of testing wrist straps daily.

Use Auditing Reports and Personal Observations to Uncover Training Problems

Review Chapter 9 on how to use an auditing report to identify problems in the ESD control program. Here is a brief review:

- Analyze the trend charts for problems.

- If a problem is not self-correcting, study the situation.
- Use the series of steps with Pareto analyses to identify the root cause of the ESD control problem.
- If the problem is a work performance issue, look for a permanent solution before immediately scheduling retraining.
- If the problem is a work performance issue because the equipment is not being used correctly, schedule further training.
- Using subsequent Auditing Reports, do a Pareto analysis of the department that received additional training to see whether the problem has been solved.

Use Engineering Solutions to Provide Permanent Solutions to Training Problems

Auditing reports and visits to the manufacturing floor will reveal work performance deviations. More training is not always the best solution. The following example explains this approach to problem solving.

It could be seen during a conversation with an employee that she was wearing her heelstrap incorrectly. She had reversed the heelstrap so that the conductive portion of the heelstrap—the part that makes contact with the floor—was on the top of her shoe (Figure 11-2B). The conductive portion should fit under the ball of the foot so that contact is made with the floor (Figure 11-2A). Also, she had attached the conductive ankle strap over her sock so that it did not touch her leg. Thus, there were two discontinuities in the path to ground from her skin surface to the floor.

Since it was highly probable that many employees were wearing the heelstrap incorrectly, the first impulse was to revise the training, emphasizing how to use a heelstrap correctly. However, a better approach was to design the heelstrap, if possible, so that it could not as easily be put on incorrectly. In this case, a design was possible (Figure 11-2C) that greatly simplified training. ***Conductive shoes offer an even better engineering solution (Figure 11-2D) because virtually everyone knows how to put on a pair of shoes***. Obviously, no additional training is needed and the original training problem is permanently solved.

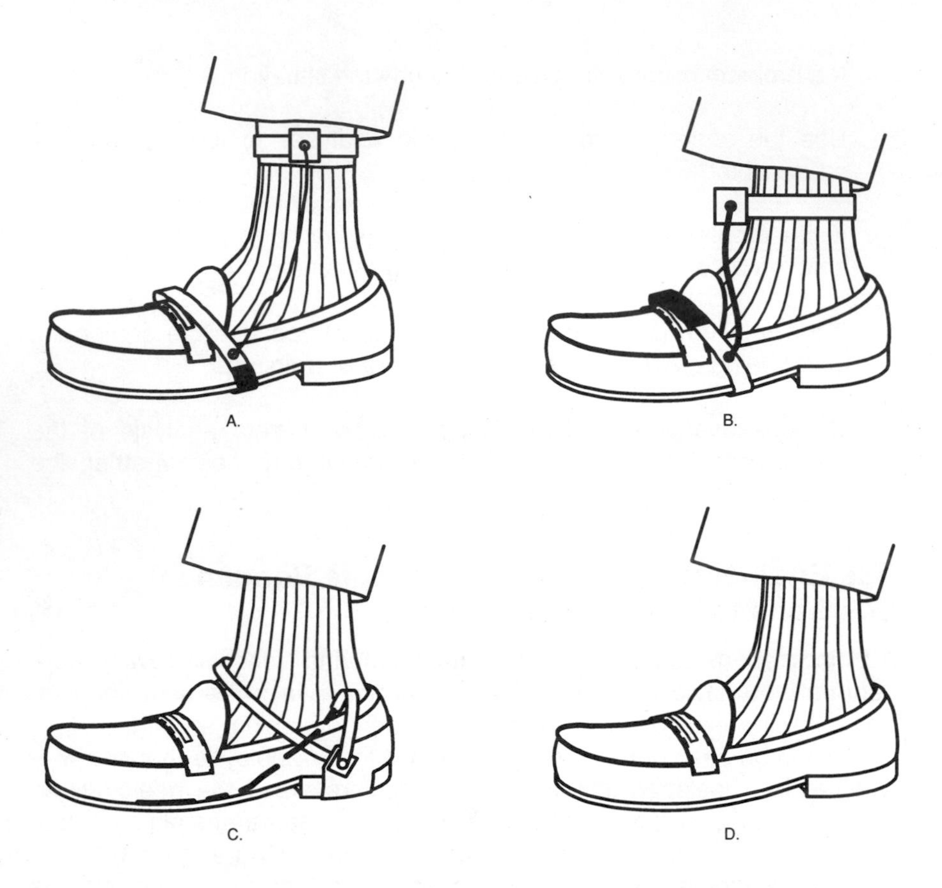

Figure 11-2. A) A heelstrap worn correctly. B) A heelstrap worn incorrectly. C) A heelstrap worn correctly and less prone to human error. D) A conductive shoe worn correctly—easiest and requires little training.

Correcting the heelstrap problem was simple compared to finding a way to train all personnel on how to handle devices of varying sensitivities. An obvious solution was to train a special cadre of employees on special techniques for handling very sensitive devices. Another possible solution was to train all employees on the different ways to handle devices of different sensitivities. This would be a complex training assignment and would require that employees decide when to use different procedures.

The solution to the varying-device-sensitivity problem was to organize work areas in the plant based on device sensitivity. Areas were designated O, I, II, III, or IV. These designations described the most sensitive device handled in that area. Each work area was equipped by the engineering staff to protect the most sensitive device. Thus, the bench employees were all trained to use whatever equipment they found at a workbench. They had no responsibility to modify their work behavior for a change in device sensitivity. The only exception to this being ultrasensitive devices which require extraordinary measures.

Thus, even though the employees understood the significance of device sensitivity, they generally did not need to change any work behavior. When assigned to an area, the work procedures remained the same but the ESD control tools varied. Training was simplified. So was the documentation.

Part Two: Planning Classes and Recommended Courses

Find Resources to Train Everybody

Everybody should attend some type of training. This includes shop employees, engineers, and all levels of management. Later in the chapter, there will be a description of courses for various groups of employees. Of course, the type of training offered will differ depending on the employees' needs and capability to learn.

With so many people needing training, difficult questions arise: How does one obtain adequate training resources? Who should attend the training sessions first? And so on. Can the program handle emergency training? To answer these serious questions, one must have a way to set priorities when allocating resources.

When forced to allocate limited training resources, use the auditing results and the Three Principles of Training. The First Principle of Training tells us to schedule training for a measurable change in work behavior. The auditing results readily identify (Figure 11-3) who will benefit the most from additional training, and what they will need to learn. Subsequent reports will indicate whether the training was successful or if follow-up training is necessary. Principle Three states that students forget information and skills that are not used regularly. Therefore, students who have no immediate need should not be scheduled for ESD training.

Auditing reports, combined with the Three Principles of Training, provide a highly effective means of allocating limited training resources.

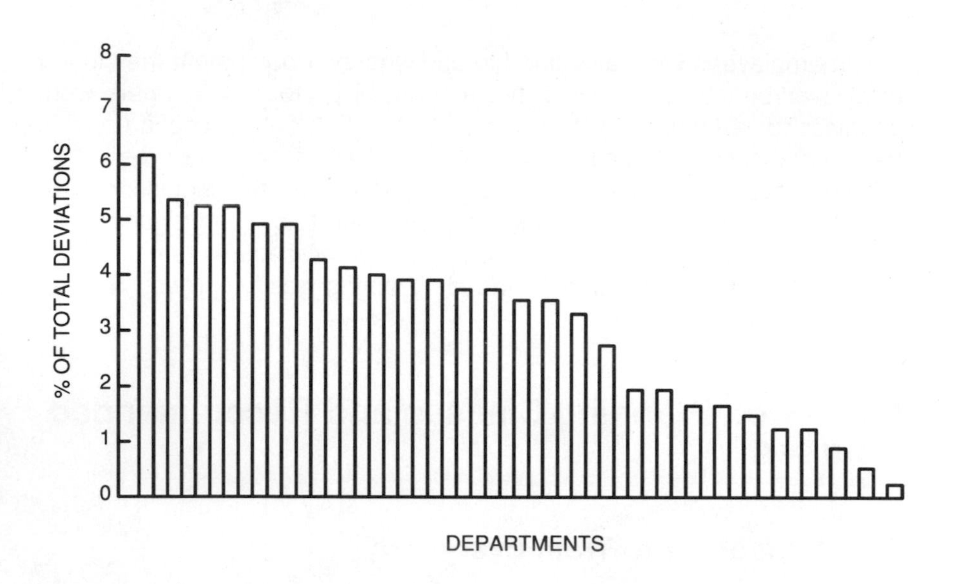

Figure 11-3. A Pareto chart of a departmental comparison readily identifies who needs training and who does not. Limited training resources can be directed to those in the greatest need.

Determine Class Size—Mass Training Versus Small Group Training

When resources are limited, it is tempting to schedule everybody into mass training. In some situations, mass training works, while in other situations, there is no alterative except to use a small group.

Use small groups if an open-ended, two-way discussion is necessary to solve a problem. This is particularly true when a group tries to reach a common understanding of subjective issues.

Small groups are also effective when a problem needs to be looked at from a team approach. For example, sessions that are populated by a vertical slice of the shop's management, engineering, and productions employees. We use a trained facilitator to encourage the group to reach a consensus on how to best comply with the Handbook. In this way, the group mounts a unified effort, and the results can be measured with the auditing program.

Ask Questions and Move About

Audiences tend to get bored when forced to listen to a person who stands behind a lectern and delivers a monotone view-graph lecture: "And now we'll discuss electrostatics...And now we'll discuss device sensitivity..."

Effective trainers learn to move about through the audience talking in a conversational tone without notes while engaging in good eye contact with the audience. They use humor; they talk with the audience, not down at them; and they ask many good questions. Quite often the right question initiates an unplanned, two-way conversation with the entire audience learning about a point they needed to know.

A good repertoire of questions should be as much a part of a trainer's teaching plan as a set of transparencies, a roomful of demonstration equipment, or a stack of training videos. Here are some sample questions to build from: Have you ever seen the plant policy on ESD? Do you know what it says? Do you know who signed it? Is it possible to tell if ESD has damaged a device? Can ESD damage a completed PWB assembly? Can you damage a device while wearing a wrist strap?

Questions that try to inform and initiate thinking build interest by encouraging participation. However, questions that make students feel stupid will only embarrass and humiliate, and consequently, discourage participation. Start with questions that require only a yes or no answer. With questions that require more knowledge, let people volunteer to answer, or call on somebody who is less apt to feel embarrassed if wrong.

Demonstrate the Nature of Electrostatics and the Effect of ESD on Devices

Electrostatic discharges are dramatic, attention-getting events and, therefore, excellent for training. There are countless numbers of demonstrations to draw from that will fit into almost any type of class. The following demonstrations are a few examples of how to demonstrate the nature of electrostatics and ESD damage to sensitive devices and assemblies.

When selecting equipment for the demonstrations, preferably use instruments with graphic displays rather than digital readouts. This is because people relate better to a line changing shape than a digital display. Furthermore, the graphs provide excellent records for further discussion and class participation. These types of instruments will enhance the level of comprehension and retention for most students. For similar reasons, analog meters are preferable to digital.

Basic Equipment: The following list of basic equipment will be useful in demonstrating the fundamentals of electrostatics and the effect of ESD on devices and assemblies. See Chapter 10 for further information on purchasing and using the equipment.

- Van de Graaff Generator
- Personnel Voltage Tester
- Personnel Charger
- Charged Plate Monitor
- Strip Chart Recorder
- Curve Tracer
- Device Test Fixture
- ZEROSTAT Pistol
- Wrist Straps (include examples of defective ones)
- Electrostatic Locator (static field meter)
- Devices of Varying Sensitivities
- Ionizer
- Faraday Cups of Varying Sizes
- Surface Resistivity Meters
- Discharge Points—Spheres and Cones
- Aluminum and PLEXIGLAS* Disks
- Sheets of Insulators
- Horsehair brush
- Insulated tweezers and pliers.

---

* Registered trademark of Rohm & Haas Company.

Basic Demonstrations

Comprehensive Exhibit—The Quality Fair: Each year at the AT&T North Andover facility, a comprehensive exhibit of demonstrations is displayed at a Quality Fair during Quality Month. All employees visit the booths set up in the exhibit hall and learn about efforts at the plant to build quality into our products.

The ESD booth has space for many exhibits and is staffed by either an engineer, trainer, or inspector who can explain the ESD control effort and answer questions. Visitors passing by usually stop to see the demonstrations in electrostatics and ESD control techniques and a continuous showing of video tapes. Most demonstrations show that a wrist strap really works. In spite of the comprehensive training program leading to the certification of most employees, we still hear comments such as "Gee, the wrist strap really works" or "That's why we wear them." The comments are gratifying.

The Van de Graaff generator (Figure 11-4) is invariably the best attention-getter.

Figure 11-4. A Van de Graaff generator will produce up to 400,000 volts, make your hair stand on end, and is the most popular demonstration at the Quality Fair.

People love to see hair stand on end and sparks fly between the two spheres. Collecting and charting voltage readings on passersby with a personnel voltage tester (Figure 11-5) also creates a lot of interest.

Another demonstration at the booth includes building a charge between two brass spheres by rubbing an aluminum disk against an insulated sheet (Figure 11-6). We also let visitors use the charge plate monitor and an electrostatic locator (Figure 11-5).

Visitors can also compare a defective IC shipping tube with a usable tube (Figure 11-7). They slide a device inside the tube and measure the charge generated on the tube using a Faraday cup. They can also see the effectiveness of an ionizer measured by the charged plate monitor or the electrostatic locator. The curve tracer (Figure 11-8) provides one of the more convincing demonstrations of device failure. People touch a proven good device and see the instantaneous failure response on the curve tracer.

A Quality Fair has no clear measurable goals. Rather, one senses that for some employees, new skills are learned, while others have a confusing point clarified.

Figure 11-5. A personnel voltage tester measures the voltage on people and is a convincing demonstration. During any one day of the Quality Fair, readings would vary from 50 volts to 8000 volts.

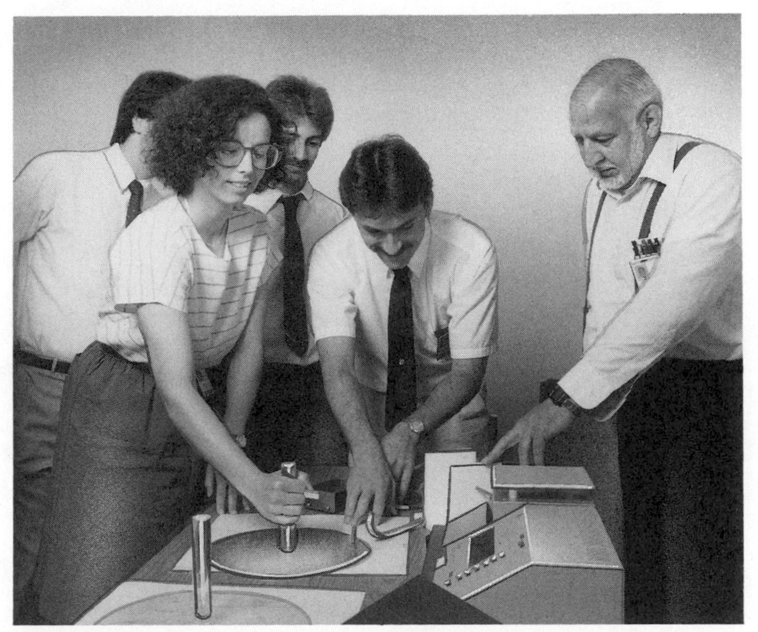

Figure 11-6. Rubbing an aluminum disk on a insulated sheet produces enough charge to create a visible discharge between two brass spheres.

However, for the vast majority, a Quality Fair continues employees' awareness of the ESD issue.

Demonstration 1: Charge Variation Among People:

Teaching Goal:

To demonstrate the wide variation in charge among people and that a wrist strap will remove charges effectively.

Equipment and Supplies Needed:

Personnel Voltage Tester
Chart to Record Voltage Readings
Wrist Straps

Ask each person who comes forward to touch the personnel voltage tester. Point to the meter reading so that the class sees from where the measurement comes. Record the readings, such as 500 volts, on the chart. Then ask that person to put on a wrist strap and touch the meter again. Record the second reading on the chart. Repeat this procedure a number of times.

Figure 11-7. A Faraday cup can be used to demonstrate that IC shipping tubes can be defective and generate damaging levels of charge on devices.

In addition to having volunteers agree to a voltage reading, increase audience participation in other ways. Appoint two laboratory assistants: one will read the meter and the other will record the readings on the chart. After the first three readings, ask the class to predict the next reading. Record their predictions on a separate chart.

The predictions will miss the mark, but the lesson is still instructive. The exercise raises the question, and hence their awareness, about which factors might affect charge generation. The first readings will show a wide variation in the voltage on people. We have found readings that varied in one class from 50 volts to 8000 volts. After the volunteers have donned a wrist strap, they will have readings of 2 volts or less.

Demonstration 2: Device Damage Without a Visible Discharge:

Teaching Goals:

To demonstrate that a person must be charged to at least 3000 volts before one can see or feel the discharge.

To demonstrate that a device can be damaged without any sensation of sight or sound.

Figure 11-8. A curve tracer provides a graphic illustration of device failure with one touch from a charged person. People relate better to a line changing shape than a digital readout from a meter.

To demonstrate that a wrist strap prevents the damage of sensitive devices even when the person is charged to 10,000 volts.

Equipment and Supplies:

Personnel Charger
Curve Tracer
Devices that Fail at Approximately 1000 Volts

Part One. The goal of Part One is to demonstrate that a person must be charged to at least 3000 volts before one can see or feel the discharge.

Select as a volunteer from the class someone that is well known, and charge the person to 1000 volts (Figure 11-9). While looking at the person and then the audience, tell the volunteer to touch ground. Ask the class and the volunteer if anybody saw or heard a spark. With the class clear that there was no visible spark, and neither did the volunteer feel any sensation, up the amount of voltage in incremental steps.

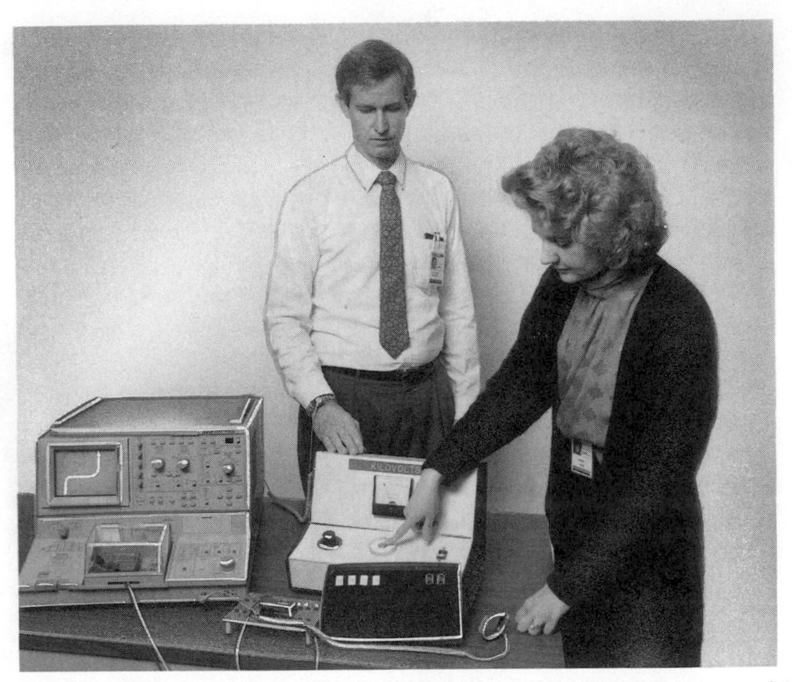

Figure 11-9. A person will not see or feel the discharge until charged in excess of 3000 volts.

With the audience's attention directed to the personnel charger, charge the person to 2000 volts and have him/her touch ground. Look for a spark and ask if he/she felt anything. Then continue the process at 3000 volts, 4000 volts, and so on. Keep raising the voltage and discharging the charge until the volunteer can feel the snap of a discharge at his/her fingertip. This usually occurs between 3000 and 5000 volts. As the voltage builds, so will the audience's interest. To have some fun after the volunteer has felt the charge, ask the audience if they would like to see the voltage raised higher. They generally take great delight in raising the voltage to the maximum of 10,000 volts to see what will happen to their colleague.

Part Two. The goal of Part Two is to build on the experience of Part One and demonstrate that a device can be damaged without any sensation of sight or sound.

Describe the I-V (current versus voltage) function on the curve tracer in understandable terms, and explain that the shape of the line is an indication of proper device function. Proceed with the demonstration only when the class has a clear understanding that any change in the line's shape means that the device has been damaged. Charge the volunteer to 1000 volts and have her touch the device. The I-V trace should show leakage on the device.

Repeat the process described above three or four times, damaging a device with different volunteers. It should become clear that a device can be damaged without any spark being seen or without the volunteer feeling the discharge.

Part Three. The goal of Part Three is to demonstrate that a wrist strap prevents the damage of sensitive devices even when the person is charged to 10,000 volts.

Using the same setup as above but with the volunteer wearing a grounded wrist strap, charge the person in incremental steps from 1000 volts to 10,000 volts. The device will neither fail, nor will the volunteer feel any sensation. Repeat the demonstration with at least two more volunteers.

Demonstration 3: Cumulative Damage:

Teaching Goal:

To demonstrate the reality of cumulative damage to sensitive devices and the reliability implications.

Equipment and Supplies:

Personnel Charger
Curve Tracer
Device Test Fixture
Devices that Fail at 1000 Volts

Using the same experimental setup as in the above demonstration, charge a volunteer to about 900 volts, just below the device's catastrophic threshold. Have the volunteer discharge through the appropriate pin of the device. The I-V trace will reveal partial junction damage. Continue the procedure of charging the volunteer to 900 volts and touching the device. Each subsequent discharge will reveal further degradation until the device fails completely.

Explain to the class that even though the device's threshold is 1000 volts, the effect of cumulative jolts of 900 volts will eventually destroy the device. Tell them how each exposure weakens the device and can result in partially damaged devices leaving the plant. These devices are less able to withstand further abuse and could fail prematurely in the hands of customers.

Demonstration 4: ESD by Induction—Device Failure:

Teaching Goal:

To demonstrate device failure due to ESD by induction.

Equipment and Supplies:

Personnel Charger
EPS Foam
Horsehair Brush
Insulated Tweezers
Curve Tracer
Device Test Fixture
Device that Fails at 1000 Volts
Electrostatic Locator.

ESD damage caused by induction is an extremely difficult concept to convey to employees and engineers alike. It is described in technical terms in Chapter 3, "Fundamentals of Electrostatics," but requires well-planned demonstrations for most people to really understand it.

This is a simple demonstration of a device failing and is done using the same setup as in the previous example, but with the addition of EPS foam, a horsehair brush, and insulating tweezers (Figure 11-10).

Charge the EPS foam with the horsehair brush, and measure the potential with the electrostatic locator. It should be charged to 5000 volts or greater. Then put a device proven to be undamaged on the EPS foam. For best results, handle the device with clean insulating tweezers to prevent leakage and drop the device onto the EPS foam to ensure maximum charge separation. Have a volunteer wear a wrist strap and touch the appropriate pin. Then lift the device from the foam with the tweezers and put it into the fixture to be tested by the curve tracer. Make sure that everybody sees that the volunteer is grounded by a wrist strap, and emphasize that in the previous experiments the device was protected by a wrist strap. Here it is not! Although this experiment may take two or three tries, the device will fail and the class will be aware of the dangers that insulators, like EPS foam, present.

Demonstration 5: ESD by Induction—Graphic Illustration:

Teaching Goals:

To demonstrate how a charge can be induced on a conductor at a distance.

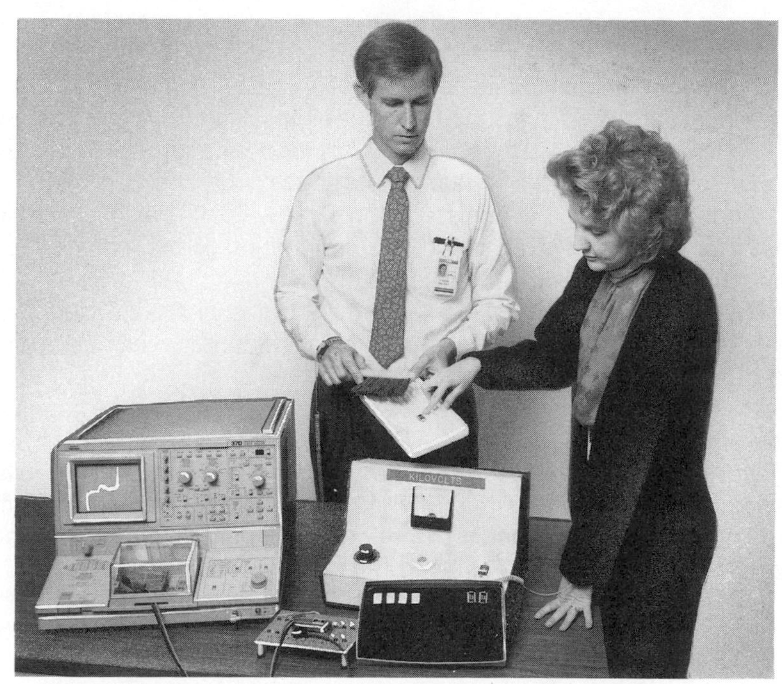

Figure 11-10. While employee is wearing a wrist strap, devices can be damaged due to ESD by induction.

To demonstrate how a device or PWB assembly charged by induction can be damaged if grounded in the presence of the field.

To demonstrate how a device or PWB assembly charged by induction and then grounded remains charged and can be discharged again—double jeopardy.

Equipment and Supplies:

Charged Plate Monitor
Strip Chart Recorder
EPS Foam
Horsehair Brush
Wrist Strap.

This demonstration is intended primarily for engineers but can be simplified for employees without a technical background. It provides a graphic illustration of the concepts presented earlier on "ESD by Induction—Device Failure."

All three of the teaching goals can be achieved while using one setup. This is accomplished by making the analogy that the plate of the monitor is a conductor and behaves the same as a device or a PWB assembly would under the same conditions.

The concepts build one upon the other in a logical way. However, each concept is difficult for some to understand, so go slowly, summarize frequently, and repeat or start again from the beginning if the class appears confused at any step along the way. It is also a good idea to have a viewgraph available of the response of the charged plate monitor for easy reference and discussion.

Adjust the charged plate monitor so that it is in the floating mode. Charge an insulator to 2000 volts, and hold it about two inches from the plate of the monitor (Figure 11-11A).

Point out to the class that the reading on the plate of the monitor is nearly equal to the charge on the EPS foam.

Move the EPS foam away from the plate, and watch how the voltage declines. Repeat, moving the EPS foam towards the plate and then away, showing that the amount of induced charge measured on the plate varies with distance. (See Figure 11-12, sections A and B).

Hold the EPS foam about two inches away from the plate, so that the reading is about 2000 volts. Have a person ground the plate by touching it (Figure 11-11C). The reading will fall to zero as shown in Figure 11-12, section C. Explain that the charge has been removed rapidly from the plate and can be simulated by the charged-device model. If the plate had been a device with a threshold below 2000 volts, it would have been damaged. This is how devices are damaged by induction.

Explain the above sequence by repeating it and using the traditional model of the alignment of charge on the two surfaces as described in Chapter 3, "Fundamentals of Electrostatics" (Figure 3–10). When a conductor (or in this case an insulator) is held two inches from the plate, the positive and negative charges will be separated. When the plate is grounded, the negative charge is removed.

Remove the ground and hold the EPS foam at the same distance away (Figure 11-11D). The apparent voltage on the monitor now reads zero. But there is actually an imbalance of charge on the plate because of the discharge. Move the EPS foam away from the plate and note how the voltage rises to a positive 2000 volts. (See Figure 11-12, sections D, E, and F). Have the volunteer ground the plate a second time. The voltage will again drop to zero and remain there. (See Figure 11-12, sections G and H). If the plate had been a device or a PWB assembly, it would have received another potentially damaging discharge. This is the double jeopardy of an induced charge.

Explain the above sequence of this second discharge by again using the traditional model of the separation of charge. When the first discharge occurred, it left a deficit of negative charge on the plate, but the charged EPS foam still maintained an induced charge on the plate giving a net zero reading. However, when the EPS foam was removed, the deficit charge on the plate was now shown to be an actual net positive charge of 2000 volts. Grounding the plate again will reduce the amount of charge on the plate at zero.

Demonstration 6: Voltage Variance With Foot Movement:

Teaching Goal:

To convey a basic understanding of voltage variations with foot movement by demonstrating that raising and lowering one's foot can change the voltage reading significantly.

Equipment:

Charge Plate Monitor
Strip Chart Recorder
EPS foam
Conductive Tote Tray.

Place the EPS foam on the floor and have a volunteer stand on it. The EPS foam is necessary to allow time to conduct the demonstration before the charge on the volunteer dissipates. It is also useful in explaining that the static voltage (V) will vary when the capacitance (C) is varied. This is defined by the equation $Q = CV$. The voltage on the person will also vary based on the charge induced by the EPS foam. The polarity and magnitude of the induced charge is primarily dependent on the interaction between the foam and the person's shoe.

Then have the volunteer hold the end of a wire that is connected with the plate of the charged plate monitor. The monitor should be in the floating mode to record the voltage response. Use a strip chart recorder to preserve the results and to graphically illustrate the effect. This will greatly enhance both the understanding and retention of this information by the students. Now have the volunteer raise his/her foot, and observe how the voltage response changes (Figure 11-13) ***significantly*** (900 volts) and directly with the foot movement. Next, have the volunteer put the foot down and the voltage will return to the starting point. Emphasize at this point, that this demonstration illustrates one of the reasons that ***continuous grounding via wrist straps or conductive footwear is so important***. Momentarily touching ground is ineffective and, therefore, not an approved method.

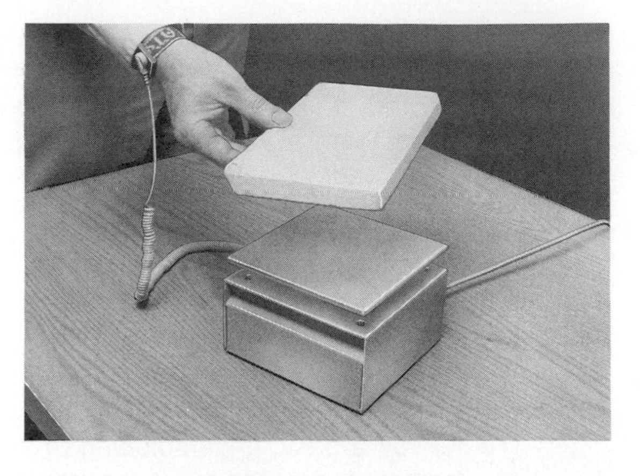

Figure 11-11A. Move the EPS foam up and down, and note the voltage variation. Then hold it steady approximately two inches above the plate of the charged plate monitor.

Figure 11-11B. Hold the EPS foam two inches above the monitor, wear a wrist strap, and touch the plate of the monitor. Note the voltage drop and discharge!

Figure 11-11C. Remove the ground and hold the EPS foam the same distance from the monitor plate. The voltage is unchanged.

Figure 11-11D. Next, move the EPS foam away from the monitor plate, and note the increase in voltage.

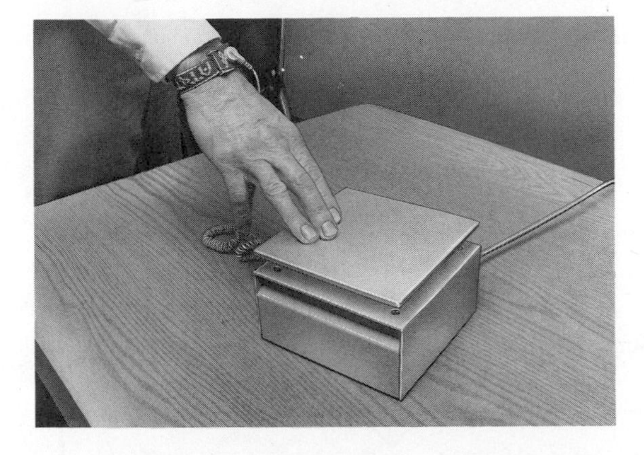

Figure 11-11E. Ground the plate a second time, and note the second discharge — double jeopardy!

Figure 11-11F. Remove the ground and note that the voltage remains at zero this time.

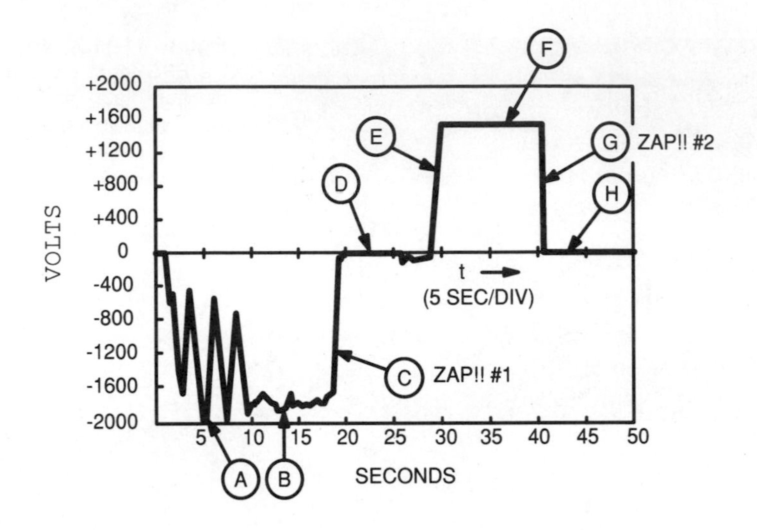

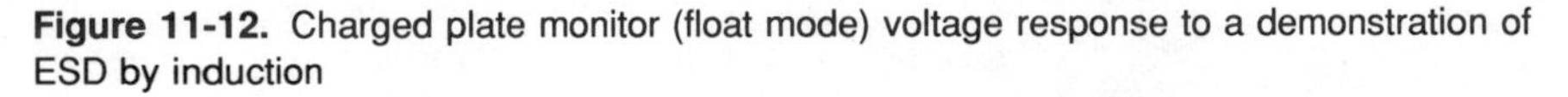

Figure 11-12. Charged plate monitor (float mode) voltage response to a demonstration of ESD by induction

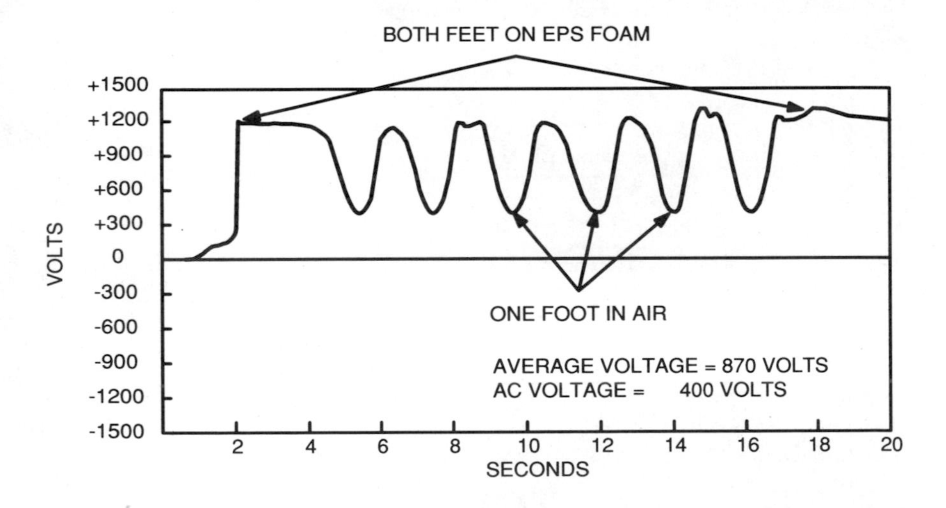

Figure 11-13. Voltage response versus foot movement. Lifting one foot resulted in a 900 volt change.

Ask the class if they know why the voltage reading changes. Explain that the induced charge varies with distance and that the person's foot acts as one plate of a capacitor. Capacitance ($C = \varepsilon A/d$) varies directly with the area of the plate and inversely with the dielectric thickness.

Therefore, the voltage reading changed because both the body capacitance and the induced charge changed.

In this instance, it can be concluded that the effects of the induced charge were greater than the capacitance variations because the voltage decreased as the foot was raised. The voltage would have increased if the change in capacitance were the controlling factor.

Voltage suppression can be demonstrated in a similar manner using the charged plate monitor and any conductive material such as a tote tray. Connect the tote tray to the monitor with a wire to record the voltage response. Charge the tray and then raise and lower it over a ground plane such as a dissipative work surface. It will be necessary to hold the tote tray with insulated pliers to prevent charge dissipation. Under these conditions, the voltage recorded by the monitor will vary with distance as it did above.

Videos: Training employees in a factory who are either entry level or have transferred to another position is a labor-intensive activity. This training is often referred to as generic skills training. To learn efficiently to perform skills such as soldering and solderless wrapping requires watching the task being performed and then practicing. Short video tapes have proven to be an excellent way for employees to learn specific but generic skills and general information about their jobs.

Many skills involved in ESD protection can be taught in a series of short tapes. These tapes about skills such as putting on a wrist strap, testing a wrist strap, handling a sensitive device, and cleaning an ionizer should be short. Four to eight minutes seems to be a functional time span to present the information adequately and fit in with scheduling other shop skills modules. Short videos have another advantage: they can be edited easily to respond to changes in equipment or rules.

Tapes can be prepared with the help of either an outside firm or a service within your company. After a topic has been selected, these people can help in preparing the script and the setting and can do the camera work. After shooting the tape, they will edit the work and, with your guidance, produce a finished product.

Recommended Courses: The following description of recommended courses reflects the need to develop and schedule courses tailored to the individual needs of many different groups of employees in the organization. The specific course material and grouping of employees will vary from one company to another. The courses should always be goal-oriented and based on the Three Principles of The Psychology of Training and Learning.

Orientation and Skills Training: An introduction to ESD control should be included in all of the company's general orientation sessions. Try to have all new employees leave the general session knowing a few introductory principles of ESD control and desiring to comply with control procedures. Of course, they will receive further instruction in their certification or specialized training.

As the new employees enter the classroom, let each one use a personnel voltage tester to record his/her amount of charge. Place the measurement on a chart, and compare the different amounts of charge carried by different people. Thus, each person should learn that the amount of charge varies considerably from one person to the next.

With the class aware of static charge, show an introductory video which should run less than thirty minutes. Be sure that the video outlines the ESD problem and describes what the company and its customers are doing to tackle the problem of ESD damage to devices.

The introductory session is a good time to hand out the ESD Control Handbook and other related printed material. Summarize the material with a view-graph talk.

This is also a good time to introduce the class to the spectacular effects of a Van de Graaff generator. This gives everybody a vivid experience to remember.

To end the session, show the new employees a few ESD control procedures that are followed at the workstations. It is particularly important to emphasize how to adjust and test wrist straps and heelstraps. Tell the employees' supervisor to do follow-up introductory training on the skills that were taught in class.

Shop Employee Certification: To certify that all employees know how to comply with all procedures described in the Handbook, we offer a three-hour training course. Each class contains about twenty people. In this course, the students will master the day-to-day work procedures on how to use wrist straps, testing equipment, handling and transporting devices, and how to identify static-generating materials. What follows is a description of the course.

When the attendees arrive, each completes a short registration form for both attendance purposes and for our certification records. When the class is assembled, they receive a very brief introduction to ESD control and are then given a pretest. The questions are either multiple-choice or true-and-false and can be completed in about fifteen minutes. The test is not signed but is handed in to be corrected by the instructor.

While the class watches a custom-made 18-minute video that answers most of the questions on the test, the instructor quickly corrects the test. The results of the tests indicate to the instructor what basic information should be stressed or reinforced in his/her teaching. After the video, the instructor leads a discussion using a view-graph and prepared overlays to teach concepts that were missed on the test. The discussion format permits the students to ask questions and to respond both verbally and visually to points that are covered.

Thus, the pretest serves two functions: the questions identify learning goals for the students and the results of the test allow the instructor to diagnose the learning needs of this particular group. Instructional goals are emphasized in the test, repeated in the video, and emphasized again in the discussion.

The students complete the three-hour certification course by taking a final test identical to the pretest. People who pass are certified for two years while those students that fail are given additional training until they pass.

As a final step in the certification of bench employees, the instructor meets with the group's manager to pass on their comments. Many times their comments are an excellent opportunity to find problem situations on the manufacturing floor that can be corrected.

There is one more optional step. Consider giving a follow-up test about six months after a training session to learn how well students have retained what they learned. At AT&T, we have found the retention to be excellent: approximately 95 percent of the students pass the follow-up test on the first try.

When the students arrive for the follow-up test, we administer the same one given at the end of the Certification course and score the results. A cut scored method is used. This means that answers to question are weighted depending on the importance of the question. We also administer a feedback questionnaire to solicit their opinions on how the class was conducted and what changes could be made in the classroom procedure or in the follow-up shop training.

Process Checker Certification: In this course, the potential process checkers will learn how to spot and correct deviations in their assigned work area. Before taking this course, they must pass the Certification course for shop employees described above.

To be certified as a process checker, the student must pass both a written test and a demonstration of proper procedures. In the demonstration, the student shows that he/she can approach a workstation, test the workbench, and record all deviations using the basic charts that are provided. Since there are only a small number of process checkers, the size of the class can be limited to twenty students. The small size means that the teaching can be responsive to the needs of a small class.

Like the Bench Employee Certification course, we start with a pretest on process checking. Names are left off of the test and the questions are true-false and multiple-choice, covering all aspects of process checking. The instructor corrects the test (to assess the needs of the class) while the class views a video that answers all of the questions asked on the test.

Using the results of the pretest, the instructor leads a detailed discussion on process checking procedures. The discussion is guided by a thirty-page Process Checking Instruction booklet that covers step-by-step procedures on process checking.

In the demonstration, the student is observed evaluating a work position where he/she accurately tests the bench and records deviations on the basic charts provided. With both tests passed, the certified process checker returns to his/her work area with the ability to detect deviations. In addition to detecting deviations, the certified process checkers become a cadre of resident experts, assigned throughout the plant, who can answer questions and offer instructions to employees and visitors.

Process Checker Charting and Analysis Training: Students attending this two-hour course learn to plot data collected during their inspections of the work area. The data entered into statistical summary charts is analyzed to learn what the defects might mean. To be admitted for this fairly selective course taught in small groups of five to ten, an employee must pass the Bench Employee and Process Checker Certification courses.

The group is guided through an analysis of a completed set of charts. Next, they are presented with the raw data from an inspection, and they complete a set of summary charts. Finally, they analyze the charts looking for trends and determining what they might mean.

Supervisor Training: This is a four-hour course taught to small groups of ten to twenty people. To be admitted to the course, the student must have passed the initial Bench Employee Certification course, the Process Checker Certification course, and the Process Checker Charting and Analysis course. When the supervisors complete this course, they will have a thorough knowledge of all ESD requirements and how to implement them. In addition, they will know what tools are available and where to get help if necessary. ***Throughout the course, it is impressed on the supervisors that they must set a good example in complying with ESD control procedures.***

The supervisors must master the information in the Handbook, the Auditor's Manual, and the Process Checker's Instruction Book. They study the data showing the economic benefits of ESD control. They learn how to explain to their employees the results of audits. They learn why there are advantages to having only one process checker per department. Finally, they are provided with a description of resources: Where to find help and sources of corrective action, including a review of construction and maintenance procedures, and a list of training courses.

Engineer Training: This course, for groups of ten to twenty engineers, lasts four hours. The engineers are given materials to study prior to attending the course. Since the physics of electrostatics and the engineering of ESD control are not subjects that are well covered in the standard engineering school curriculum, their instruction here should build upon a basic knowledge of electrical theory and the general practice of engineering.

The goals of the course are to teach engineers how problems get solved in ESD control engineering, how to detect problems, and what measures to take after they detect a problem. Thus they will learn about the physics of electrostatics, the comprehensive nature of an ESD control program, such as how to use measurements to aggressively pursue detecting problems, and the resources to be used in solving problems.

Prior to attending our course, each engineer studies the AT&T video tape and attends a basic certification course about the requirements of the ESD program. In addition, he/she studies a PC-based self-instruction program on the fundamentals of ESD, case studies of past problems that have been solved, and ESD control implementation techniques.

During the four-hour class, the engineers review the certification course and learn how compliance is enforced. ***Since engineers tend not to be faithful about compliance, its importance is stressed.*** They learn not only how noncompliance damages devices, but also the implications of their noncompliance. We quote what employees, who must wear wrist straps all day, say about engineers' attitudes regarding compliance. We also mention that compliance at AT&T is a condition of employment.

After a review of the basic principles of electrostatics, case studies in ESD control problems are analyzed with the basic principles of ESD control engineering emphasized. Chapter 3, "Fundamentals of Electrostatics," makes a good reference for this course. Many of the examples on ESD control engineering in this book could be used for engineering-training purposes.

Maintenance and Construction Personnel Training: This is a three-hour course taught to groups of ten to twenty people. A twenty-page Installation Standard For ESD Control Items is used as a study reference. During the course, the trainees install and test grounding hardware. They also study how to install and test workbenches, different types of tabletop materials, and floor mats.

Awareness Training: There are a number of other techniques that help to keep employees aware of the ESD issue. Announcements on the plant's public address system about changes in ESD procedures or reminders to the entire plant of items such as the need to test wrist straps and heelstraps daily are common awareness techniques.

Producing and distributing booklets to illustrate topics such as safe work conditions or the proper ways for an employee to be equipped for the best ESD protection can also bring dividends. Cartoon figures are very useful for making these types of illustrations. Publish periodic bulletins for manufacturing management and engineers to describe changes in ESD practices and equipment or to bring ESD practices into line with changes in the manufacturing process.

A yearly letter about ESD practices from upper management to the entire work force reinforces much of a coordinator's training efforts. The letter should be an assertive statement about the company's policy and why ESD practices are necessary. This will urge management commitment and employee compliance by reminding the entire work population that ESD protection is essential.

Periodically reissuing the Handbook also helps make employees aware of ESD control. Since the program has continuous improvement, changes are inevitable. When the Handbook is changed, publish the new editions of corporate and local handbooks for the supervisors and engineers.

Last of all, hold periodic meetings with shop and engineering organizations. At these meetings, report on the progress of projects and manufacturing efforts. These meetings are an ideal way to keep employees aware of the ESD control effort.

Conclusion

A training program built on measurable work performance goals allows the coordinator to aggressively pursue the identification of, and solution to problems. This means that the ideal training program would teach a work force so well that there would be zero deviations or perfect compliance with the ESD program requirements. The measurable goals should shape the Handbook and dictate what is taught. These same goals should appear as items on the Auditor's Checklist.

The training goals should be reached by using the best possible teaching methods. This includes discussions, demonstrations, and videos that motivate the learner with the content and presentation shaped to his/her needs. The training should be comprehensive, directed at changing the work habits and attitudes of an entire culture. Well prepared courses should reach the entire work force with follow-up training whenever necessary. Last of all, make use of every possibility, such as Quality Fairs, public address systems, bulletins, and booklets, to keep everybody aware of the ESD control effort.

Points To Remember

- Plan the training program around measurable goals that teach all employees how to comply with the ESD control procedures described in the Handbook.
- Make the ultimate goal of the training program to achieve zero deviations in work performance or perfect compliance with the ESD program requirements.
- Study the Three Principles of The Psychology of Training and Learning.

  Principle One: Train only to affect a measurable change in work behavior.

  Principle Two: Motivate students to improve learning.

  Principle Three: Take into consideration that students forget information and skills that they don't use regularly.

- Prepare measurable goals and the course outline based on the Handbook and the Auditor's Checklist.

- Look for problems in the ESD control program by observing work behavior and analyzing auditing reports.

- Motivate students with instruction that is purposeful and interesting.

- Develop good questioning techniques, well-executed demonstrations, and carefully prepared short videos to keep interest high.

- Motivate learners with pretests, which also gives the instructor a diagnostic tool.

- Use a Van de Graaff generator, a personnel voltage tester, and a charged plate monitor with a curve tracer as basic equipment for ESD demonstrations.

- Use a Quality Fair as an interesting way to teach ESD awareness and the importance of compliance to a large number of employees in a short period of time.

- Include a brief introduction to ESD in the New Employee Orientation Meeting.

- Tailor each course to the needs of each functional level or class of employees. Their jobs will define their needs.

- Use all available means, such as the public address system, bulletins, booklets, revised versions of the Handbook, and letters from management, to provide a steady flow of information about ESD control and to keep the entire work population constantly aware the issue.

Chapter 12

Packaging Considerations

Although packaging is intended to provide protection for its contents, it can be the cause of ESD failures unless special materials and/or procedures are used. The successful implementation of ESD-protective packaging procedures depends on having a basic understanding of how sensitive devices may be damaged in a package or during the packaging procedure and how the protective procedures work.

In Chapter 3, "Fundamentals of Electrostatics," we discussed the two basic events that can ultimately lead to the destruction of microelectronic devices: charging (triboelectric or contact) and discharging. We know that both of these events could, ideally, be avoided by preventing the motion inherent in the triboelectric charging process, by minimizing contact between or with insulators, and by keeping all surfaces at equal potentials. In practice, however, this is not possible. Electronic manufacturing by its very nature is a constant blur of motion. Devices must be moved from place to place and come in contact with a variety of materials. As a result, highly visible controls, such as the wrist strap, still leave ESD sensitive device in considerable jeopardy. In the absence of room air ionization, an expensive and controversial approach to static control, the battle lines must be shifted to the selection of static control materials and the deployment of special ESD control.

In this chapter, we will present the basic concepts to consider in using special materials for ESD-protective packaging and surfaces. These concepts apply not only to "traditional" packaging (cartons, bags, and boxes), but also to temporary packaging used during manufacturing (tote boxes) and to other surfaces that may contact a device during manufacture (bench tops, rails).

Since the introduction of ESD control materials in the late 1960s, some terms have been invented to describe these materials according to the way that they attack the problem. The precise definition of these terms has changed slightly over the years.

The following paragraphs are based on definitions in Chapter 7, the Glossary section of this book, and on the EOS/ESD Association Glossary of Terms. This information has been repeated here for convenience and expanded upon as it pertains to packaging considerations.

- **Antistatic:** Materials that effectively prevent the build-up of a static charge on themselves or on materials with which they come in contact.

- **Static dissipative:** Materials that slow down the otherwise extremely fast discharge involved in a CDM event. The EOS/ESD Association and the Electronic Industries Association define these as materials having surface resistivities between 10^5 and 10^{12} ohms/square.

These two material types directly address the charging and discharging steps involved in most failure scenarios. Their use, in combination with some other simple strategies, has been shown to provide for the broad protection of sensitive devices, even in modern automated assembly factories. In some cases, however, another set of materials is brought into play.

- **Conductive:** Materials that are used in various applications to maintain the leads of a device at the same potential. They have a surface resistivity of less than 10^5 ohms/square. Depending on how they are used, conductive materials may also be referred to as screening or shielding materials. These terms generally apply when the material is intended to be used to prevent field-induced dielectric breakdown through the attenuation of external electrostatic fields. Experience has shown that this type of protection is only necessary for a few very sensitive metal oxide semiconductor (MOS) devices that employ no on-chip protection. This type of shielding should not be confused with the electromagnetic interference (EMI) shielding of operating electronic equipment. While effective electrostatic shielding can be obtained with materials of relatively high resistivity, 10^9 ohms/square in some cases, effective EMI shields typically employ materials with resistivity below 10 ohms/square.

Even today, there is considerable confusion over the meanings of these terms. Many materials may be both antistatic and static dissipative. Furthermore, it is quite common for conductive materials to generate a charge on some insulators. That is, they are not antistatic. Understanding these distinctions and how they apply to specific situations is critical to implementing and maintaining an effective control program and to evaluating vendor claims about the effectiveness of their products. All of these material types must also have properties that will not interfere with the normal course of manufacturing. Abrasion resistance, thermal stability, contamination effects, and many other properties may be important elements of the overall material specification.

Antistatic Materials

As we saw in Chapter 3, insulators become charged by contact with other materials. During the contact a number of physical processes occur that allow charges (electrons or molecular ions) to flow across the boundary between the materials. An antistatic material minimizes this charge flow. When we refer to a material as antistatic, it is not a complete description. This is because the tendency to tribocharge is actually the property of two materials or objects. A more complete description would be that a given material is antistatic with respect to another. In practice, the other materials are insulators (such as an epoxy/glass PWB substrate) or conductors (such as copper traces on the PWB) that may be charged in a given process or handling procedure. Commercially successful antistatic materials are those that are antistatic with respect to a large portion of the materials encountered in their usual applications, and it is these materials that tend to get the generic label antistatic.[53]

There are three types of commercial antistatic materials:

1. Treated with a special agent called a topical antistat.
2. Synthetic polymers that are impregnated with an antistatic agent which is insoluble in the polymer and "blooms" to the surface.
3. Intrinsically antistatic.

Topical antistats are particularly useful because they minimize charging between many otherwise widely different materials. They usually consist of a carrier or solvent and the active antistat. Some examples are quaternary ammonium compounds, amines, glycols, and amides of lauric acid. Application of an antistat introduces a layer between materials that tends to dominate the interfacial properties. The mechanism by which these antistats, also called surfactants, reduce tribocharging is not completely understood. It is known that surfactants are hygroscopic, i. e., they promote the adsorption of water at the surface. In fact, their function is usually highly dependent on the ambient relative humidity. They also reduce friction. This effectively reduces the total time of intimate contact between the materials as well as the number of contact/separation cycles. The accompanying reduction of frictional heating may also be important. Furthermore, since the antistats are, at least at moderate humidities, somewhat conducting, they may dissipate or spread some of the charge that is transferred. Although this latter property may be useful, it should not be regarded as necessary or sufficient for an effective antistatic material. Antistatic materials should be able to perform their intended function, the reduction of charge generation, without being grounded.

Material Characterization

A number of test methods have been proposed for evaluating antistatic materials. A particularly popular one is the Inclined Plane Test first introduced by Huntsman[54] (Figure 12-1). In this test, standard objects made of TEFLON* material or quartz are allowed to slide down a plane covered by the material of interest at a fixed angle. The charge transfer to this sliding object is measured using a Faraday cup. This conceptually simple test is typical of the philosophy of the other tests that have been developed.

---

* Registered trademark of E. I. Du Pont de Nemours & Co., Inc.

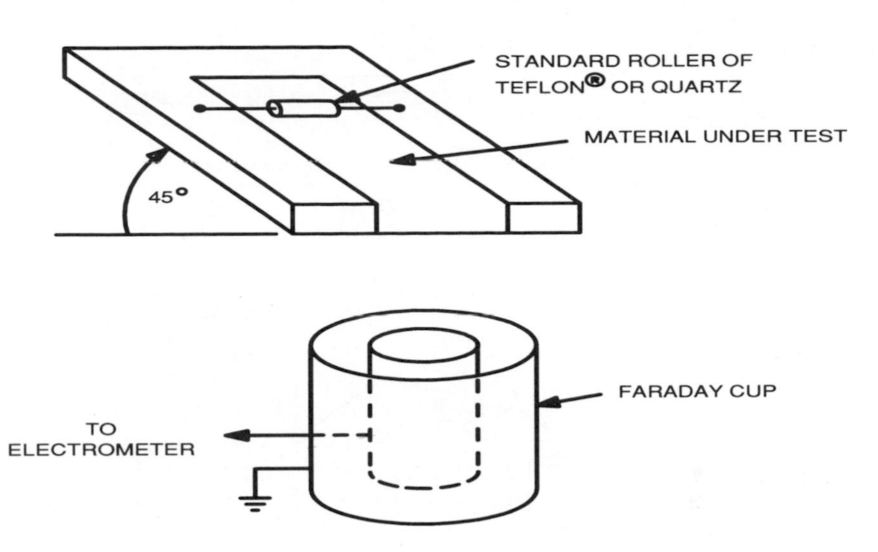

Figure 12-1. Schematic of Huntsman's Inclined Plane method

No one standard has emerged because of the difficulty in correlating the results of the simple test to actual processes. Instead, special tests for specific products have been proposed. Some examples are tests for IC shipping tubes,[55] bags,[56] and tape-and-reel packaging.[57]

These tests simulate the motions that are associated with a specific process. In the shipping tube test (Figure 12-2), an integrated circuit is allowed to slide several times along the length of a tube before it is allowed to fall into the cup. In the tape-and-reel test, a device is vibrated in its carrier tape and is charged by its repeated contact with the cover tape (Figure 12-3). Because of problems with cover tape adhesion, most cover tapes are insulating on the side facing the device.[57]

Surface resistivity is not a good measure of the relative antistatic effectiveness of different materials. However, variation in surface resistivity can be a good indicator of variation in the material's tribocharging propensity. Once this correlation has been established, the relatively simple resistance measurement can be used as an acceptance test for that material. Relative humidity should be controlled and recorded when testing antistatic materials.

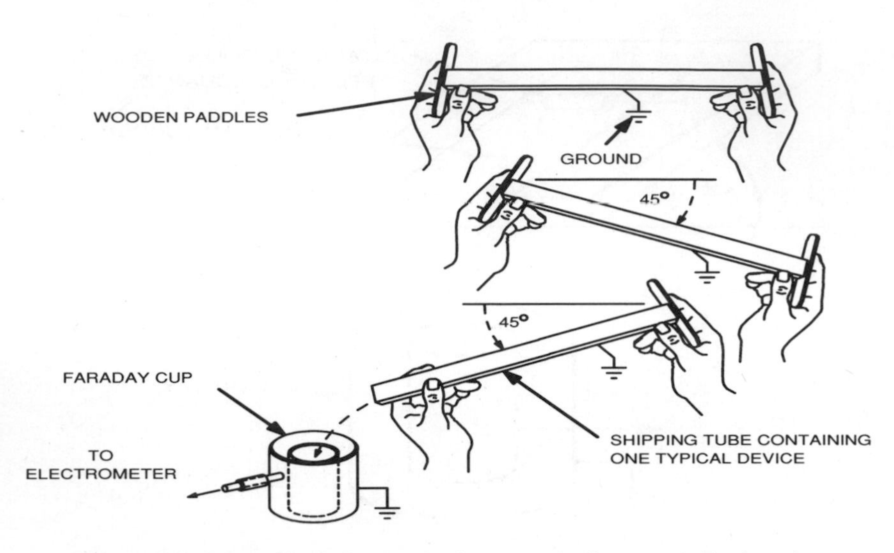

Figure 12-2. Schematic of triboelectric charge generation test of shipping tubes

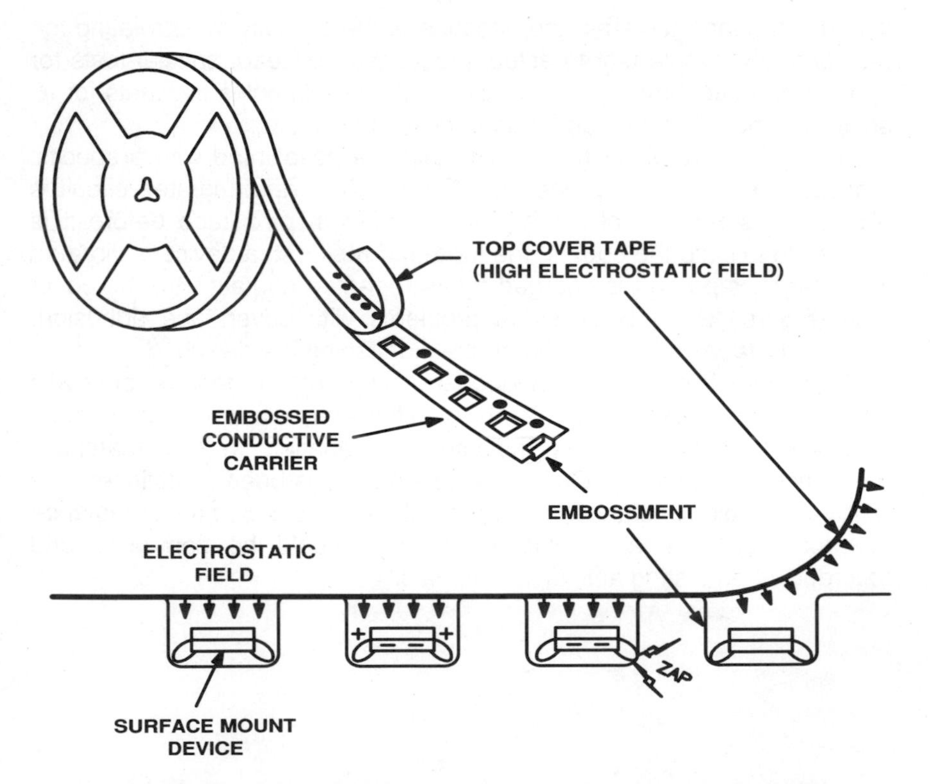

Figure 12-3. Surface-mount device packaging (tape-and-reel) field-induced failure

Static Dissipative Materials

Since there are many instances where charge generation cannot be avoided, these charges must be safely removed. Many antistatic materials can also function in a static dissipative manner when grounded or employed in large sheets (as in flooring). Other static dissipative materials may be homogeneous volume resistive or may be laminates with a conductive core, as for bench tops. The term "static dissipative" was actually invented to describe a class of materials which limit the current that flows through a charged device when it comes in contact with the surface. The EIA and the EOS/ESD Association have defined this relatively vague property as any material that has surface resistivity between 10^5 and 10^{12} ohms/square. Bossard et al.[29] has shown that the 10^5 ohms/square lower limit is appropriate for protecting energy-sensitive devices that obey a specific thermal model for device failure.

Material Characterization

The most direct indication of the effectiveness of a static dissipative material is the resistance between any point on the surface and ground. In a fixed installation such as a workbench, this measurement is the most practical one because it probes not only the material but the ground connection as well. However, it is often necessary to separate the material properties from those of the installation. The EOS/ESD Association standard test method for work surfaces calls for the following resistance measurements:

1. Between points on the surface,
2. Between a point on the surface and a groundable point (see Chapter 6, Instrument #8).

However, for other static dissipative items, the situation is not as simple. Irregular objects such as tote boxes and IC shipping tubes (Figure 12-4B) are generally characterized by any resistance measurement that is practical.

Another method that is used is the static decay method. This test evaluates the ability of a static dissipative material to allow the removal of a charge from itself when it is grounded. In the test, an item is charged to a known potential (for example, 2000 volts), and the decay of this potential is monitored as a function of time.

This decay should follow the exponential decay predicted for an RC circuit:

$$V(t) = V_0 e^{-t/T}$$

Where T = RC is the time constant. For a tote box for PWB assemblies, the capacitance is about 50 pF. In a typical specification, the potential is required to decay to a fixed percentage of its original value, say 1 percent, within a specified minimum time, say 2 seconds. Thus, for our 50 pF tote box, we must have:

$$R < \frac{(t/c)}{\ln(V_0/V(t))} = \frac{(2/5\times10^{-11})}{\ln(100)}$$

$$= 8.7\times10^{9}\ ohms$$

which is in the middle of the static dissipative range. Again, as with antistatic materials, relative humidity is an important factor and should be controlled and recorded during both the resistance and static decay test.

Conductive Materials

Materials with surface resistivity below about 10^5 ohm/square have been given the name conductive. They may be used to assist in the removal of charges from other conductors or static dissipative items (such as a tote box on a conductive work surface) and to maintain common potentials. The most common application is the familiar black conductive foam used for packaging small numbers of sensitive devices. Considerable care should be exercised when these materials are used because of the possibility of CDM damage.

Conductive materials used for ESD applications are usually either polymeric materials loaded with some form of carbon particles, such as the foam mentioned above, or metallized (laminated or vapor-deposited) structures such as those used in some ESD-protective bags.

While 10^5 ohms/square has been established as a boundary between static-dissipative and conductive materials, it should not be regarded as the lower bound for safety from CDM damage. For example, the conductive materials that are used as carrier tapes in the tape-and-reel packaging of surface-mounted integrated circuits are probably safe down to about 10^4 ohm/square. Knowledge of these variations can be of great practical significance when the only materials available are several conductive materials in the range 10—10^4 ohms/square (which is currently the case).

Material Characterization

Clearly, a critical property of a conductive materials are their relatively low resistance, which establishes a reliable ground. Thus, the general methods applied to static dissipative materials should also apply. However, some resistance-measuring equipment may not span the full range 10—10^{12} ohm/square (see Chapter 6).

In cases where the shielding effectiveness of a conductive material is important, several other tests have been proposed. Among these are the capacitive probe test,[29] the voltage monitoring test,[58] and various field attenuation tests.[29]

Capacitive Coupling and Air Gaps

One of the reasons that ***conductive (shielding) materials are seldom necessary*** is that the orientation and position of a device with respect to the source of a static charge can be sufficiently restricted to keep any detrimental effects to a minimum level. Some examples that utilize air gaps to accomplish this are shown in Figure 12-4 and are discussed in the following sections.

Integrated Circuit Shipping Tubes

The potential that a device in an IC shipping tube is exposed to is shown in Figure 12-5. Here C_C is the capacitance between the source and the device and C_D is the capacitance of the device. V_S is the electrostatic voltage of the source. The voltage seen by the device is then given by:

$$V_D = \frac{V_S C_C}{C_C + C_D}$$

That is, the rigid shape of the tube guarantees that there will be a gap, which will reduce V_D. Unger[55] has argued that the ratio V_D/V_S is typically about 1/50. It follows that as long as external voltages are kept below about 5000 volts, devices with 100 volt thresholds (for dielectric breakdown) would not be in jeopardy. Conductive or metallic IC shipping tubes appear not to be necessary unless the device is extremely sensitive and high voltages are likely. In fact, Unger has also shown that conductive tubes are more likely to transfer charge to a device because they allow charge to distribute rapidly over the length of a tube.

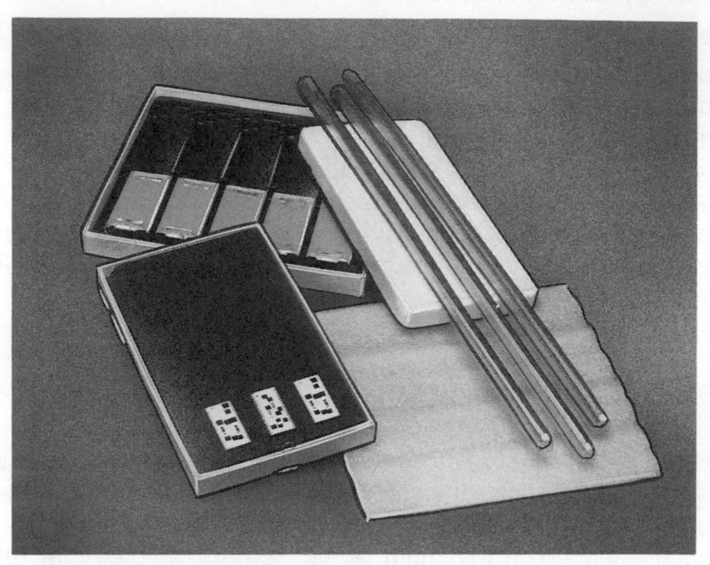

A. IC shipping tubes and static dissipative foam in boxes

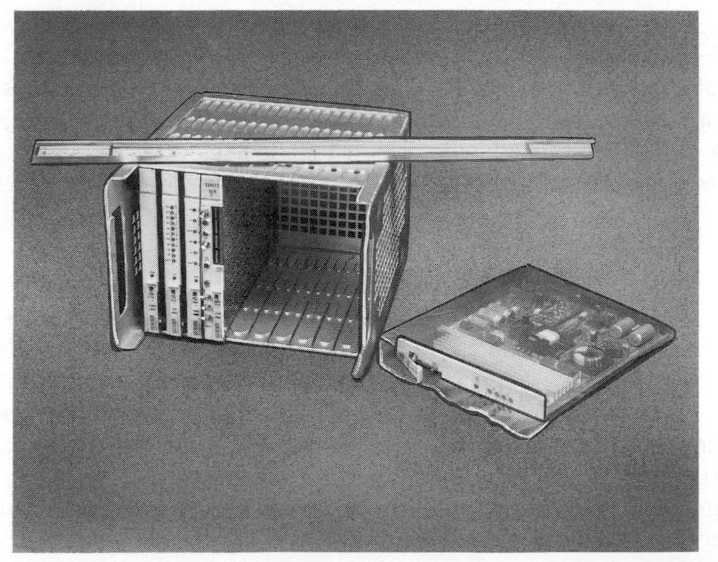

B. Static dissipative anodized aluminum tote box and IC shipping tube and static dissipative bag without shielding

Figure 12-4. Examples of packaging that utilize air gaps

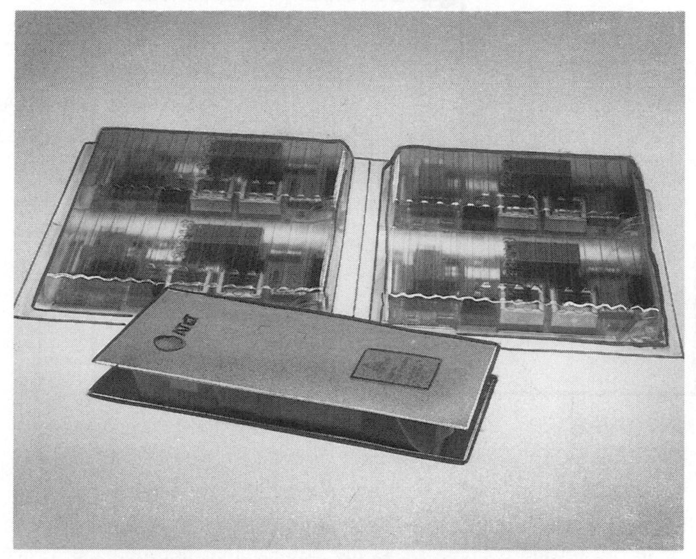

C. Rigid topically treated bubble pack (or blister pack)

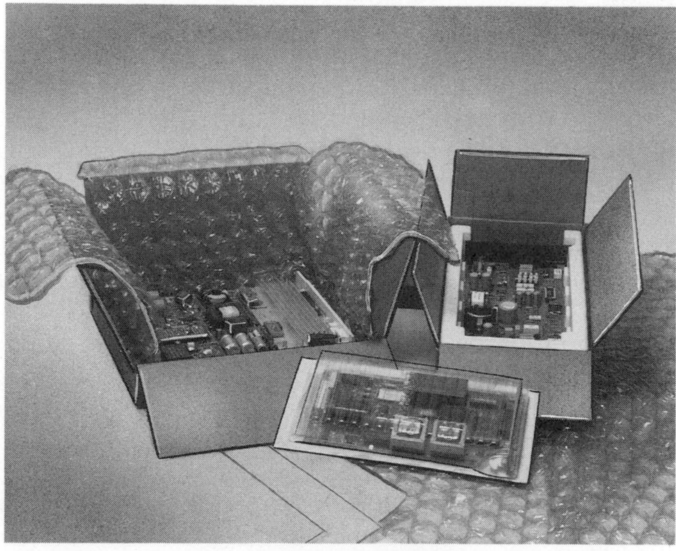

D. Antistatic bubble wrap

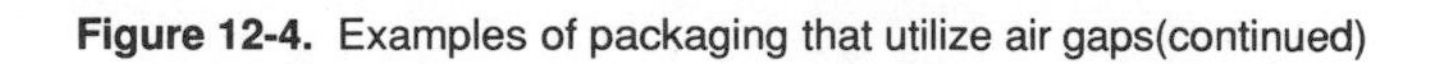

Figure 12-4. Examples of packaging that utilize air gaps(continued)

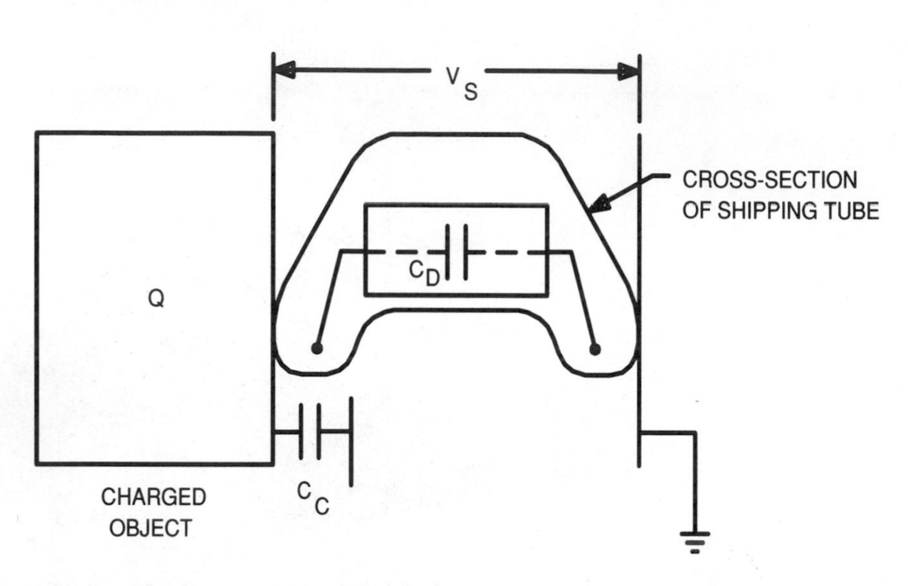

Figure 12-5. Coupling of a device in an IC shipping tube to an externally charged source. V_S is the potential on the source. C_C is the capacitance between the device and the source, and C_D is the capacitance of the device to ground.

Tote Boxes

A typical anodized aluminum dissipative tote box used for circuit boards is shown in Figure 12-4B. (Anodized aluminum can be specified to have a surface resistivity between 10^6 and 10^{10} ohms per square.) Most tote boxes are made of a static dissipative material so that any charge that might appear on them may be removed by grounding or by placing them on a static dissipative or conductive work surface. Still a charge may appear and remain on an outer surface or on an object near the box. Again, the orientation and position of the boards minimizes coupling to these sources. This is illustrated schematically in Figure 12-6.

The surfaces of the box that are in contact with the board have a relatively weak coupling because they are perpendicular to the plane of the board. On the other hand, the surfaces that are parallel have reduced coupling because they are required to be about 1/2 inch away because of the positioning of the slots in the box. In terms of the parallel plate capacitor model (Chapter 3), the perpendicular surfaces minimize the exposed area of the plates while the parallel surfaces have a maximum distance between the plates.

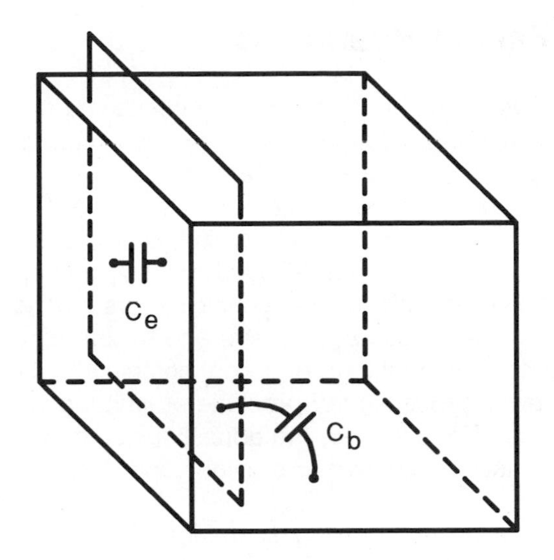

Figure 12-6. Schematic of the coupling of a PWB assembly to the bottom, C_b, and parallel edge, C_e, of a tote box.

The amount of additional protection afforded by this type of arrangement is difficult to quantify. However, compared to the coupling seen by a circuit board that is mounted horizontally over a plane charged source, typical reductions in coupling are about 1/2. Since sliding the box is the most common source of tribocharging, charges will most often reside on the bottom surface. Therefore, the orientation of the circuit boards in Figure 12-4B is forgiving.

Bubble Packs Versus Bags

When a rigid or semirigid material is used, air gaps can be maintained during shipping and handling that simultaneously provide visibility and physical isolation and that minimize field effects without resorting to conductive materials. Flexible bags, however, can be forced into "worst-case" coupling scenarios where charged sources can be brought to within a few thousandths of an inch of a sensitive item. Indeed, most of the studies that have been done on electrostatic shielding have been done on bags in worst-case configurations with extremely high source voltages (15,000-35,000 volts).[54,59-61] In most realistic cases, minimizing coupling through orientation, combined with other protection procedures in this chapter, is sufficient to keep ESD losses to a minimum.

Fighting ESD Failure Scenarios

The merits of the various protection materials and packaging procedures can be illustrated by walking through the three most common ESD failure scenarios. These are summarized in Table 12-1.

Each of the scenarios involves a few identifiable steps where protection techniques can be applied. Since each process is sequential, presumably one could eliminate the threat by eliminating any one of the steps. In practice, however, this is not possible. First, none of the techniques are foolproof. Antistatic agents age and become ineffective. Ground contacts become intermittent or open completely. Secondly, other objectives of a given process may preclude or limit the application of some of the remedies. For example, the adhesive side of some tape-and-reel cover tapes must be insulated to provide good bonding with the carrier tape.

The steps referenced in the following scenarios are listed in Table 12-1.

Scenario A: Two surfaces (1) experience some motion that produces a static charge, (2) then a sensitive device is placed in the field of this charge, and (3) the device is subsequently grounded.

Step 1 can be minimized by using antistatic material(s) while these surfaces are in contact. Any charge that remains after Step 1 can either be allowed to dissipate by using a static dissipative material, neutralized by using air ionization, or suppressed by using air gaps or electrostatic shielding. Finally, the effect of Step 3 can be minimized by ensuring that any discharge is controlled and slow (using a static dissipative material).

Scenario B: In Scenario A, there are a number of means of minimizing the probability of the final step occurring. This is to be contrasted with Scenario B (Table 12-1). Here, a static charge appears on a device because of contact with another surface.

As in Scenario A, Step 1 can be counteracted by using an antistatic agent on the contacting surface. Of course, the device cannot be made antistatic or static dissipative and still function properly. The ceramic or plastic body of the device must be highly insulated to meet electrical requirements, as well as moisture and corrosion resistance requirements. This also limits the response to Step 2. Since the charge now resides on the device package, the only way to remove the charge (other than wait for an extended period) is through air ionization. Step 3 is again the failure step. The only viable means of protection is to avoid contact with conductors and to discharge the device through a static dissipative material.

Table 12-1. Implementation of ESD Controls in Three Common ESD Scenarios

Scenario	Step	Remedy
A	1. Motion produces a charge on surface(s)	• Antistatic material or agent on either surface
	2. Devices moved near a charged surface	• Dissipative surface • Suppression by air gaps and/or shielding • Air ionization
	3. Device is grounded while near a charged surface (CDM)	• Static dissipation
B	1. Motion produces a charge on the insulated package (lid) of device	• Antistatic material or agent on contacting surface
	2. Charge remains on the device	• Air ionization
	3. Device is grounded while charged (CDM)	• Static dissipation
C	1. Charge generated by the movement of a person	
	2. Charge remains on the person	• Grounded wrist straps • Conductive or static dissipative floors and shoes • Room air ionization
	3. Charge is transferred to device by touch	• Isolation of the device • Static dissipative package slowly discharged • Conductive shunt

A discharge through a lumped resistor of 10^5-10^{12} ohms is not a good substitute for a large sheet of material such as a work surface[62] because discrete resistors have parasitic (shunt) capacitances that allow a significant flow of current at high frequency (as is seen in typical CDM ESD events).

Scenarios A and B are likely to occur because of the movement, often automated, of devices through a manufacturing operation. In this context, the familiar human threat is not a significant factor. It has been our experience that these scenarios have the greatest potential for damage because they can produce static charges in a systematic fashion as part a manufacturing process.

Scenario C: Because of the relative efficiency of personnel grounding in controlled manufacturing areas, the human body model threat tends to be a sporadic one. However, wrist straps will fail and some employees will occasionally fail to use them properly, even in a well-audited program. Thus, packaging must provide some additional protection from personnel discharge in the factory. This protection is even more important outside the factory when a circuit board may be in a somewhat less controlled repair or maintenance environment.

Scenario C describes the interaction of a charged person with a device and the protection alternatives provided by various packaging schemes (Table 12-1). In Step 1, the charge appears on the person. To prevent this, which is analogous to Scenarios A and B, eliminate the charging mechanism. This is seldom actually done. Even the most "antistatic" carpeting is actually static dissipative or conductive. The wrist strap actually applies to Step 2, removal of the charge from the individual. Another alternative is room air ionization, an expensive, incomplete, and seldom applied solution. Step 3 is the transfer of the charge to the device, either by removing the board or device from its package or by touching the package directly. ***Assuming that dielectric breakdown of the device is not an issue at this stage, we need***:

1. Adequate insulation from the charged source so that no rapid discharge occurs

2. Sufficient conductivity (static dissipation) so that, when the source approaches, any discharge will slowly leak onto the package surface.

Most antistatic dissipative bag materials are sufficient. For very sensitive devices (below 100 volts), more caution may be necessary. Again, ample additional protection may be provided by using rigid materials to ensure a significant air gap and a reduced coupling with external charged sources.

In going through these scenarios, it is clear that several opportunities are available for greatly reducing the threat of ESD failures by using antistatic and static dissipative materials. These materials directly address the critical steps towards failure. It is also clear that a considerable amount of choice is available in building an ESD control program. This point often gets lost when dealing with vendors of ESD control materials and products. One of the primary benefits of understanding the relative importance and effectiveness of the various options is being able to critically evaluate vendor claims.

For example, the program described in this book minimizes the use of conductive and shielding materials. ***When confronted with claims that conductive materials are necessary, the following points are helpful***:

1. Devices that fail simply in the presence of a field (due to internal dielectric breakdown) rather than by ESD are extremely rare.
2. Electrostatic fields can be dealt with by using air gaps and subsequent minimal capacitive coupling.
3. Arcing directly through a static dissipative bat without shielding is unlikely below 5000 volts.
4. Published studies where shielding effectiveness is demonstrated often involve extreme worst-case scenarios.
5. An increase in conductors in the environment can increase device vulnerability to the CDM; that is, more conductive is not necessarily better.
6. Using conductive materials for "the most sensitive devices" cannot be implemented in a rational way if MIL STD HBM thresholds are the only available data. This is because:
    a. The voltage thresholds for unterminated devices are much higher than MIL STD data (in the MIL STD test, devices are grounded).
    b. HBM data do not correlate with dielectric breakdown sensitivity.
    c. Selective implementation of shielding to prevent ESD due to static induction requires CDM data.

7. Experience with programs such as those described in this book have been very successful without resorting to the broad use of shielding materials (see Chapter 14).

Points To Remember

- Static dissipative materials provide sufficient protection for the vast majority of packaging applications. However, they should also be antistatic.
- Antistatic materials prevent tribocharging and should be selected with a specific application in mind.
- Static dissipative materials allow for the slow and safe removal of charges from sensitive items.
- In general, all packaging materials, as well as personnel, must be grounded before the contents are handled.
- An air gap should be included in packaging designs wherever possible.
- The properties of ESD control materials can vary significantly and should be monitored when the material comes into the shop as well as periodically during use.
- Conductive and shielding materials are generally not necessary and should be reserved for ultrasensitive devices with thresholds below 200 volts.
- Conductive materials introduce an increased risk of CDM failures for two reasons. First, they may not be antistatic and could charge devices. Second, if the material is too conductive, a charged device coming in contact will experience a CDM discharge.
- Devices that fail simply in the presence of a field (dielectric breakdown) are extremely rare.
- HBM threshold data do not correlate with field strength or dielectric breakdown sensitivity. For instance, most devices with an HBM threshold of 200 volts will not fail when simply exposed to a 10,000 volt field.

Chapter 13

Automation

The problems surrounding automation-caused ESD damage differ considerably from those typically encountered in the manufacturing process. First, the damage done by automated systems is usually because of a charged device instead of a human body. Secondly, the repetitious nature of automation can systematically damage or, worse yet, partially damage large quantities of devices, and thereby create a serious quality and reliability problem. The third major difference between automation damage and the typical ESD problem concerns the solution: more engineering skill is needed to eliminate, or at least minimize, automation-caused ESD damage.

After examining the nature of automation-caused ESD damage, this chapter offers a two-part solution. Part one explains how to implement and manage a program for qualifying automated equipment in the factory. Part two describes case studies where automation was discovered to be the culprit and then eliminated as a problem. These cases will help an ESD coordinator develop problem-solving strategies for tackling automation-caused ESD damage to devices and assemblies.

The Nature Of Automation-Caused ESD Damage: Different, Serious, and Difficult

A build up of charge on either a device or a machine is inevitable with the repetitious movement of automated machine parts in contact with devices or PWB assemblies. A high degree of charge and the highly conductive machines are the ideal conditions for a rapid discharge of energy that can either damage or destroy devices by means of the charged-device model. Clearly, automation offers conditions for causing ESD damage, and the evidence shows that it can be a significant problem.

First, there is the evidence that automation might be the culprit. When device failure analysis reveals the telltale footprints of device failure due to the charged-device model, there is a strong chance that a piece of automated equipment caused the damage. Another indication that the ESD damage was caused by automation is a sudden decline in the yield of a robust product. When automation causes ESD damage, there is a strong possibility for total and systematic damage to almost every device, or assembly, in a job lot. This scenario has been known to have yields approaching zero, similar to the situation described in Chapter 4 in which the mass soldering machine caused damage to resistors in an automated process.

One advantage to automation-caused ESD damage is that when the source of the problem is found and the problem corrected, the yields will return to normal indefinitely.

In a perverse sense, it is fortunate when the damaged devices or assemblies actually fail within the plant. The assemblies can be reworked and the problem solved before shipment to customers. A more serious issue is the likelihood of systematically wounding a large number of devices below their threshold. This can result in latent damage. In other cases, the devices suffer cumulative damage. Whether the damage is latent or cumulative, the company will ship these devices as they pass the final test, but they may become inoperable later on, thus causing customer dissatisfaction.

Automation can either damage or partially damage devices on a scale of far greater magnitude than any problem at the workstation. Furthermore, we expect this problem to grow as automation is used more frequently. Add to this the continuing trend towards using devices of ever-increasing susceptibility, and the problem grows significantly larger. These are very good reasons why a program to reduce automation-caused ESD damage should be implemented.

Many companies have not implemented this program. In fact, the most serious side of the automation-caused ESD damage issue is that the problem is frequently overlooked. Most plants have the typical ESD control protection, such as wrist straps, grounded mats, ionizers, and so on, but no provisions to prevent machine-caused damage.

The ESD control industry seems to have shied away from tackling automation-caused ESD damage. Why? Probably because the solution is far more complicated. Solving the automation problem requires technical knowledge about the nature of ESD events and the use of unusual test equipment as well as strong engineering problem-solving skills. The information that follows will permit the ESD coordinator to meet the problem head-on.

Part One: Prevention Considerations and Techniques

Automation-caused ESD damage must be controlled from a manufacturing and design point of view. It is not enough to purchase "antistatic equipment" from suppliers. Often companies supplying manufacturing equipment are not aware of and do not know how to solve ESD problems. Many suppliers do not have the necessary test equipment or don't know how to conduct the necessary tests. The final decision on whether equipment will damage devices must come from you.

The most effective way to control automation-caused ESD damage is to have a sound technological understanding before the automation process is begun. ESD problems are, as a whole, easier to deal with prior to the design of the manufacturing equipment. (See Case Study 6, Page 274.) It is difficult to change fully developed facilities; it is relatively easy to change development plans. In some cases, it is even impossible to institute adequate safeguards after the equipment is fully developed.

Qualification Criteria and Test Equipment

The first step in making a facility safe from ESD damage is to develop a set of acceptance criteria for equipment used in automation and to gather the necessary test equipment (Table 13-1). The qualification criteria in Table 13-2 will be useful in understanding and implementing the qualification process.

Table 13-1. Basic Test Equipment to Test and Qualify Manufacturing Equipment for ESD-Safe Operation
Surface Resistivity Meter
Resistance-to-Ground Meter
Ohmmeter
Electrometer
Faraday Cups (for devices and PWB assemblies)
Electrostatic Voltmeter
Electrostatic Locator with probe
Charged Plate Monitor
ESD Simulators (tripod and hand-held models)

Table 13-2. ESD Qualification Criteria for Manufacturing, Testing, and Handling Equipment Used in Automation
1. All conductive materials must be grounded.
2. Movement must not cause excessive triboelectric charging on PWB assemblies or devices.
3. All surfaces in contact with PWB assemblies or components must be static dissipative or conductive per EIA 541.
4. The manufacturing facility shall not be subject to ESD damage or malfunction due to ESD.
5. The manufacturing facility shall be equipped with provisions for the convenient wrist strap grounding of personnel.
6. Work surface shall be covered with a grounded, Engineering approved, static dissipative mat or tabletop laminate.

Criteria 1: All conductive materials must be grounded. A floating metal object presents three hazards that can easily be avoided with proper grounding.

First, an ***ungrounded conductive object can carry a substantial charge*** and thus jeopardize the product being handled. (See Case Study 1, Page 267.) If a sensitive device or assembly comes in contact with the charged object, a rapid discharge will occur that can result in device failure.

Second, ***the charged object can also jeopardize the electronics in the automated manufacturing equipment***. Depending on where the discharge occurs, the resulting transient can cause device damage within the manufacturing equipment or cause a temporary malfunction (see Criteria 4) as a result of the radiated field associated with the discharge.

Third, ***an ungrounded object is not able to effectively remove a charge from the sensitive product***. If the metal is floating and in contact with the sensitive product, excessive triboelectrification of the product can result (see Case Studies 1 and 2, Pages 266-269). Therefore, when the charged device or assembly comes in contact with the next grounded surface, a rapid discharge will occur, which can also cause device damage. This is simulated by the charged-device model.

These three hazards can be easily avoided by ensuring that all of the conductive elements of the manufacturing equipment are grounded. This can usually be done using a standard ohmmeter to verify continuity with a proven ground such as an equipment ground. A limit of less than 1 ohm to ground is sufficient.

However, a number of commonly used surfaces, such as anodized aluminum, exhibit high surface resistivities (see Criteria 3). In these cases, a resistance-to-ground meter will be necessary to verify continuity, and the 1-ohm limit no longer applies. The reading will depend on the material properties, but should be less than 10^{10} ohms.

Criteria 2: Movement must not cause excessive triboelectric charging on PWB assemblies or devices. Although totally eliminating the charge on products that move through automated equipment is almost impossible, the level of charge should be measured and compared to the susceptibility of the product. For instance, a device that fails at 50 volts should not be subjected to an automated process that consistently produces a potential of 500 volts.

Testing for this can be involved and is most accurately done using a Faraday cup and an electrometer to measure the total charge on the sensitive device or assembly. This can only be done if the product can be removed from the piece of equipment and placed in the Faraday cup without removing or adding to its charge. (See Case Study 5, Page 273.) Insulated pliers and large Faraday cups are useful for this procedure, especially when PWB assemblies are involved. However, this procedure is often not possible due to the nature of automated equipment, and therefore, other means of measurement will become necessary.

The static potential can often be measured directly with an electrostatic voltmeter. However, be aware of ground planes for they can dramatically influence the readings. Take care to ensure that there are no ground planes immediately behind the item being tested. The small spot size of the voltmeter helps.

Correcting this problem often requires proper grounding or substituting different materials for items such as conveyor belts. (See Case Studies 1 and 3, Pages 267 and 269.)

Criteria 3: All surfaces in contact with PWB assemblies or components must be static dissipative or conductive per EIA 541. A number of materials are commonly used in automated equipment, and some of them pose a significant risk of damage. TEFLON* is one such material. It is ideally suited to automation in all respects except for its triboelectric properties (see the material on triboelectrification in Chapter 3). Therefore, uninformed designers frequently use it, and (as a result) introduce ESD problems that can shut down a product line. These materials must be avoided whenever possible.

All surfaces that actually come in contact with a sensitive product must be either static dissipative or conductive. (See Case Study 2, Page 268.) This makes it possible to ground them and control the degree of charging.

A dissipative surface is preferable because it serves as a safeguard against the charged-device model. It should be used anywhere that an employee might place a device as well as on the internal portions of the equipment that routinely contact sensitive products.

Anodized aluminum is a cost-effective solution for many internal applications. The anodic coating can be regulated to produce surface resistivities between 10^6 and 10^{10} ohms per square.

Criteria 4: The facility shall not be subject to ESD damage or malfunction due to ESD. Most automated equipment includes electronic controllers and microprocessors that are subject to ESD damage or malfunction, which is usually overlooked. ESD can damage test sets and cause excessive downtime (see Case Study 4, Page 271) or can cause momentary malfunction and, thereby, falsely indict good product (see Case Study 3, Page 269).

---

* Registered trademark of E. I. Du Pont de Nemours & Co., Inc.

When an air discharge occurs, a transient is produced and intense fields are radiated. Both can adversely effect electronically controlled equipment and must be taken into account. They can ***result in undesired logic shifts that produce anomalous operation of the equipment or test set***. These responses are usually self-correcting and difficult to diagnose without an ESD simulator.

ESD simulators produce a controlled transient that can be used as a diagnostic tool to solve a problem or as part of the initial qualification. A variety of simulators exist that can either be mounted on a tripod or hand held. The testing procedures are detailed in IEC Publication Number 801-2 and are intended for the design qualification of commercial electronic equipment. Repeated discharges are introduced at all points of likely hand contact.

Criteria 5: The manufacturing facility shall be equipped with provisions for the convenient wrist strap grounding of personnel. Provisions for convenient wrist strap grounding should be installed on new equipment by the supplier and specified in the purchase order. (See Case Study 5, Page 273.) This will ensure proper installation and is more cost-effective than retrofitting the equipment after it arrives at the plant. It will also prevent any unnecessary delays in using the equipment.

A banana jack is the most common grounding receptacle, and it should be installed in a manner consistent with the plant standards. It will need to be tested for proper grounding during the initial qualification as well as periodically after installation in the facility. It should also be clearly labeled for employee convenience and safety.

Criteria 6: Work surfaces shall be covered with a grounded, Engineering approved, static dissipative mat or tabletop laminate. The purchase order should specify a specific type of dissipative work surface material so that it will be consistent with the rest of the facility. Left to the discretion of a supplier, it is likely that an unapproved surface will be installed and need to be resurfaced to conform to factory standards. Furthermore, unless a static dissipative work surface is specified, the supplier will almost invariably provide an unapproved non-conductive work surface because of the lower cost. Informed suppliers may ask which surface is needed, but this should not be left to chance.

Initial qualification should include a complete test of the work surface as well as proper grounding. This will avoid any unnecessary rework or delays and ensure product safety.

Qualification Procedures

After the qualification criteria are determined, the location for testing new equipment must also be determined. If inspection resources are limited, it is best to do an in-plant inspection when the equipment arrives. Often the equipment is not fully assembled until it arrives at its final destination, and therefore, an in-plant evaluation is necessary. This happens when separate elements are manufactured in different locations and then assembled at one location.

Either way, some elements of qualification can be done at the supplier's plant. This will help to recognize problems early and avoid unnecessary delays or rework. A final qualification procedure should always be performed at your facility. After you deal with suppliers long enough to have confidence in their quality control and test techniques, it may only be necessary to do periodic sampling.

Determine whether your qualification procedure will be formal or informal. Often in small operations, an informal procedure is sufficient. In a larger company, a formal qualification procedure is necessary to track if and when a supplier and piece of equipment have been qualified. A label identifying qualified equipment is an inexpensive, straightforward record-keeping system for any size plant. This label should identify that the equipment has passed the certification tests and when the tests took place. This system of record-keeping has an added benefit: It allows an ESD inspector to report on the status of all equipment during a program audit.

Automated equipment must be requalified periodically. To determine a schedule, resources and experience must be considered. Requalification might be necessary annually or semiannually for some equipment, while shorter or longer time intervals may be necessary for other equipment. Any mechanical change that might affect a piece of automated equipment is a signal that requalification is necessary.

Some equipment, such as IC handlers, once qualified, may only need continuity checks to verify that the grounds are connected and are effectively shunting the charge to ground. These checks should be included in the scheduled, periodic maintenance of the equipment. Other, more complex equipment, such as a mass soldering machine, requires scheduled, periodic reevaluation by the ESD coordinator or some other well-qualified person.

Part Two: Automation Case Studies

Although automation-caused ESD damage is predictable, it takes experience to recognize the problems and identify solutions. The following examples identify specific automation-caused ESD damage problems and explain the procedures used to solve them. The examples should also be seen as guides to identify ways of thinking about and solving automation-caused ESD problems.

Case Study 1: An Ungrounded Conveyor. A premium price was paid for a conveyor thought to be antistatic and conductive. (See Figure 13-1.) The product travels across black conductive rollers that were specified in the purchase order. They are mounted on a metal shaft and are driven from below.

Initial qualification uncovered a problem. The equipment was received with the metal shafts suspended in insulated bushings and consequently not grounded. It behaved like a Van de Graaff generator, delivering a potential that measured, on the boards, as high as 3000 to 4000 volts.

Figure 13-1. A conveyor with conductive rollers that were not grounded due to insulated bushings

The problem was corrected by substituting conductive bushings. Now the boards pass over the equipment with a typical potential of less than 50 volts.

Testing the board can be done in a number of ways. A static locator is the simplest way to test, but can give misleading results. The close proximity to the ground plane will produce erroneous readings due to voltage suppression. Suspending the board in air would give a better reading of the static potential resident on the board.

Removing the board from the conveyor is the preferred way to perform the test. Handle the board using a pair of insulated pliers to minimize charge transfer off the board to the handling tool. Then place the board in a Faraday cup to measure the total charge on it, both mobile and immobile.

In summary, this conveyor initially failed Qualification Criteria 1 and 2 (Pages 262 and 263). It was ***not grounded properly*** and as a result, it was charging the boards excessively. This condition could have led to significant yield problems had it not been for the initial qualification. Therefore, by qualifying the conveyor prior to use in production, a potentially serious problem was avoided. In retrospect, a simple continuity test between the shaft and ground would have been sufficient to detect and correct the problem.

Case Study 2: Faulty Grounds on a Device Handler. It has been common knowledge for many years that DIP handlers can produce ESD damage to devices. Triboelectrification can occur when the devices slide out of IC shipping tubes and down the channels of the handler. The charged devices then contact grounded conductors and suffer CDM damage.

The widely accepted solution is to use antistatic materials for the channels and to ground all of the conductive portions of the handler (Figure 13-2). Suppliers almost invariably specify continuity tests as part of the periodic maintenance because ground wires can break or become loose.

Despite the widespread knowledge of the potential DIP handler problem, initial qualification of a handler revealed a number of missing ground connections. This could have led to countless numbers of damaged or, worse yet, wounded devices. It is also why suppliers specify verification of all ground connections. Continuity tests must be done both initially and periodically to ensure proper grounding.

The initial qualification of handlers should also include Faraday cup test procedures similar to those for IC shipping tube testing. The cup can be positioned at the bottom of the channels to catch the devices as they exit the channels. The total charge on the device can be measured

Figure 13-2. A DIP handler with numerous ground connections that need to be verified both initially and periodically.

this way and then compared to device thresholds. Once a handler has satisfied this qualification procedure, the triboelectric properties are not likely to change significantly. Therefore, it does not need to be included as part of the standard periodic requalification. It should, however, be reverified if problems with the handlers are suspected.

In summary, this case study reinforces the need to ***qualify all equipment regardless of prior history***. Even the best of suppliers can make mistakes, and therefore, nothing can be left to chance. As with the previous example, the DIP handler failed Qualification Criteria 1 (Page 262); but this time, it was due to a number of missing ground connections instead of a design flaw. ***Faulty or missing grounds are a common failing of automated equipment and should be tested.***

Case Study 3: False Test Indictment. The test set in Figure 13-3 is old, with a long history of reliable performance. A problem occurred when a conveyorized system was introduced to transport products in tote trays to and from the test set. Once the conveyor belt was installed, the circuit boards being tested experienced a 33 percent failure rate. The repair mill was inundated, and we had a manufacturing crisis on our hands.

Figure 13-3. The introduction of a conveyorized transport system caused a reliable test set to malfunction and falsely indict a good product.

A detailed investigation showed that the boards failing the test were actually acceptable. The test set was incorrectly indicting a good product. Then, we found that the test set, although old, was not defective. By simply shutting off the conveyor belt, most boards passed the test. When the belt was turned on, the test set once again failed one in three boards. A strong field from the conveyor was causing the test set to falsely indict the product. (See Criteria 4, Page 264.)

When the conveyor was first installed, the belt was nonconductive and highly prone to charging. We got average readings of 6000 volts on points along the belt where there was a metal plane behind the belt. Readings of up to a staggering 35,000 volts were obtained where the belt existed in a free space with no grounded metal plane behind it. It was this field and the associated discharges that were causing the test set to malfunction.

The problem was solved by substituting a conveyor-belt material that was conductive and less prone to charging. We now measure potentials of less than 2000 volts on the belt, which is below the threshold of where the test set experiences disruptions.

In summary, it is important to note that, in this case, although it was the test set that malfunctioned, ***it was the introduction of automation that caused it***. Therefore, as automation is introduced into existing product lines, it is necessary to requalify existing test sets and equipment.

If the qualification procedures had been in place when this test set was originally purchased, the set would have failed Criteria 4 (Page 264). At that time it might have been possible to redesign the test set to eliminate its sensitivity to fields. In this instance, replacing the conveyor belt was the most sensible solution.

Case Study 4: Component Damage to a Test Set. A bed-of-nails test set used to test bare PWB assemblies (Figure 13-4) was out of commission or down 15 out of every 20 days due to ESD-caused damage to the components inside the test set. The problem occurred when the PWB assemblies, which are naturally prone to charging, were placed in the test set under a vacuum. (The vacuum and associated movement of the boards were part of the charging mechanism.) The resulting transient had sufficient energy to damage the internal electronics of the set and put the set out of service.

Figure 13-4. A bed-of-nails test set for bare PWB assemblies was frequently damaged by ESD until air ionization and wrist straps were introduced.

An analysis revealed that the input junctions of the devices in the switch cards were wired directly to the test pins in the bed-of-nails. The threshold of these devices was 1000 volts HBM, and the same threshold was determined for the test set. This was established by using a hand-held ESD simulator as shown in Figure 13-5. Frequently the boards being tested measured in excess of 4000 volts. The resulting discharge from the board through the pins to the switch card in the set clearly exceeded the threshold capability of the test set.

The situation was further complicated when the test set maintenance department replaced the damaged components with new components that also turned out to be defective due to ESD damage. These components had been stored in an IC shipping tube that had been reused so often that the topical antistatic treatment on the tube had worn off. As a result, approximately 25 percent of the replacement components were damaged by static as they were removed from the IC shipping tube.

Typically, a topically coated shipping tube is effective for only one year or one pass through a complete manufacturing process. Reuse of shipping tubes should be kept to a minimum.

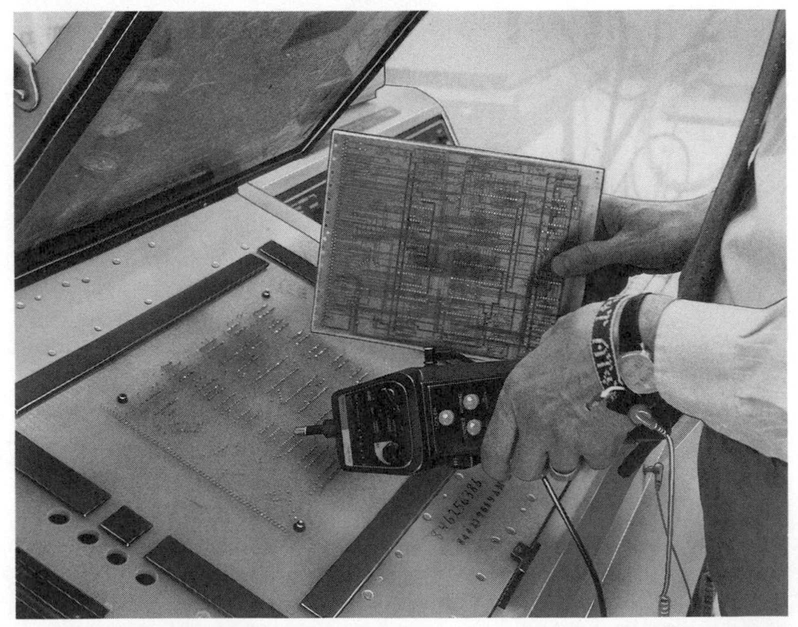

Figure 13-5. A hand-held ESD simulator being used to establish the ESD threshold of a bed-of-nails test set

To prevent any reoccurrence of the problem, this area was reclassified from Class IV to Class I. It is now considered one of the more sensitive areas in the building due ***entirely to the test sets***, not the product. Many precautions are associated with a Class I designation such as wrist straps, dissipative work surfaces, conductive flooring, and dissipative tote trays.

Despite these precautions, the problem persisted, so we added a portable bench ionizer (positioned as in Figure 13-4) that would neutralize the bare PWB assemblies prior to being tested. This solved most of the problem. With the combination of an ionizer, wrist straps, dissipation mats, and trained personnel, we now experience minimal downtime.

In summary, if the test set had been qualified using Criteria 4 (Page 264), the problem could have been avoided or at least identified prior to production. This is another good example of why it is necessary to test manufacturing equipment for susceptibility to ESD damage or malfunction per Criteria 4.

Case Study 5: Employee Discomfort. An employee at a defluxing station refused to work. At this operation, the reverse or solder side of PWB assemblies are cleaned by means of a rotating nylon brush that is partially submerged in cleaning solvent (Figure 13-6). As the employee moved the board over the rotating brush, the charge steadily rose to almost 8000 volts until an arc of almost one-quarter of an inch long occurred between the employee's knee and the bench. At the adjacent work position, another employee doing the same job and working on identical equipment was apparently untouched by any painful static discharge.

The second employee was observed resting his stomach against the grounded metal surface of the work position. His stomach formed a path to ground, bleeding away the charge as fast as it accumulated. This led to the short-term solution of requiring all employees at the defluxing station to wear wrist straps. However, provisions for convenient grounding of wrist straps had to be installed because this problem occurred prior to the general use of wrist straps.

Although this solved the problem initially, an even more effective solution was discovered shortly thereafter. When an electrometer and a Faraday cup were used to study the problem, charge accumulation in excess of 0.2 nanocoulombs per square inch was measured on the PWB assemblies. This was unacceptable, and it was therefore decided that the cleaning solvent should be replaced with a more conductive solvent to bleed off any charge generated by the rotating brush.

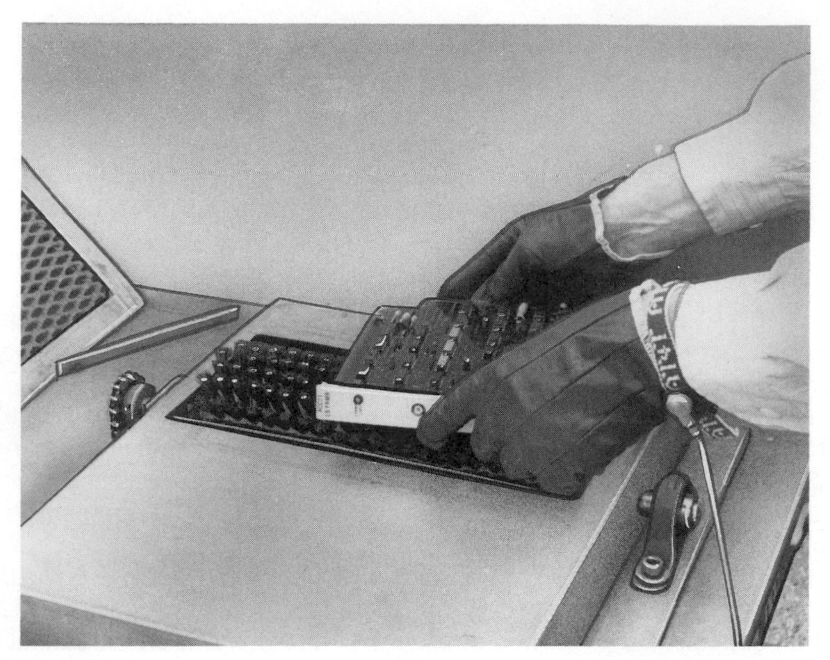

Figure 13-6. A rotating nylon brush at a defluxing station can generate up to 8000 volts without proper grounding.

In summary, the problems encountered with this operation were originally detected due to employee discomfort and preceded the general use of ESD control procedures. Today this operation would have failed Criteria 2 (Page 263) because of the excessive charging as well as Criteria 5 (Page 265) because of the lack of convenient provisions for wrist strap grounding.

Case Study 6: Preventing Charge Separation by Design. A cover is put on hybrid integrated circuits (HIC) to give dimensional stability for automated handling. Initially this presented a dichotomy of requirements in the design of an automated cover assembly machine (Figure 13-7).

Because the cover is in contact with the leads of the device and any conductivity would interfere with circuit functionality, it had to be a good insulator. But if the cover was a good insulator, it would be prone to holding a charge and thereby introducing the risk of ESD damage. The question then was how to design an automated cover assembly machine to deal with this dichotomy.

This was accomplished by designing the equipment so that all of the leads of the hybrid would automatically be grounded prior to putting the cover in place. Thus, the field that is present on the cover during

Figure 13-7. An automated hybrid integrated circuit cover assembly machine with vibratory bowl for feeding covers into the machine

assembly could not cause any charge separation or risk of ESD damage. However, to be sure, it was necessary to test the equipment.

The configuration of the equipment made it virtually impossible to test the HIC in place. Therefore, it was decided to have the device fall into a Faraday cup, but this too caused problems: the cup could not be positioned at the exit of the equipment. The solution was to reshape the exit tube (Figure 13-8) to permit the hybrid to fall into the cup directly. The tube was heated so that it could be bent into the desired shape. Of course, due to the heating, the tube had to be retreated with a topical antistat and retested to be sure of the triboelectric characteristics.

In summary, this example illustrates the benefit of ***designing automated equipment with ESD in mind***. When the equipment is nothing more than a concept, it is relatively easy to include ESD protection. After the fact, the necessary changes can be costly and in some cases impossible.

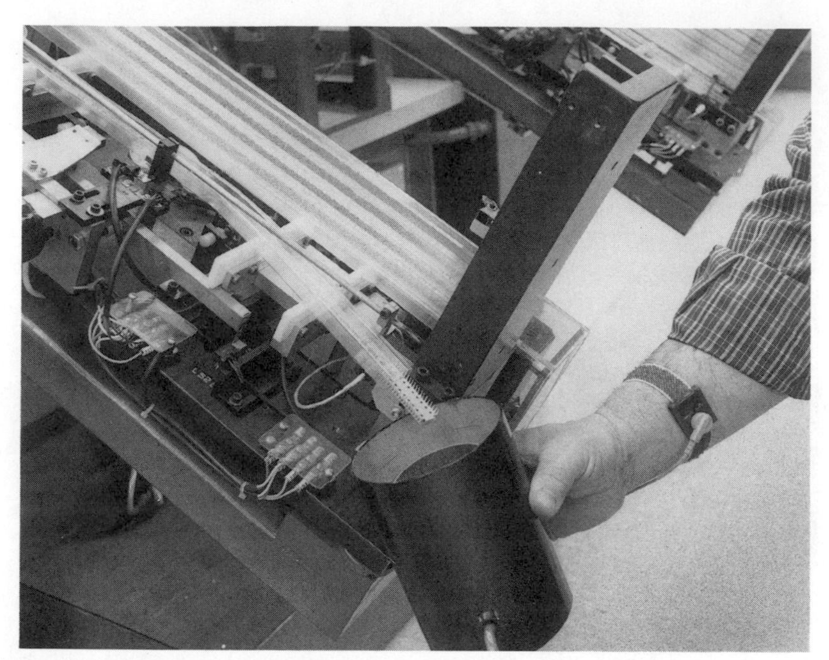

Figure 13-8. Faraday cup testing of an automated hybrid integrated circuit cover assembly machine

Conclusion

As many companies become skilled at preventing ESD damage to devices from people, the dominant mode of failure shifts to that of automated equipment. Not only can automation cause many more failed devices than the random failures induced by people, it is a problem that is often overlooked. Nevertheless, the problem can be solved by understanding the charged-device model, which is the dominant failure mode associated with automation, and including the qualification of the manufacturing facility as a vital part of an ESD control program.

Using the examples and advice in this chapter, tailor your plan to suit your facility and resources. Be sure to qualify all equipment. Follow the "buyer beware" rule at all times. Many suppliers do not understand the physics of ESD damage, nor do they understand the magnitude of the risks involved. Also, many do not understand how to test the equipment. Assume that much of the equipment you buy will produce ESD-caused damage to devices. Then, place your best efforts in the design phase of your facility. Once the equipment is assembled, a design change is almost impossible.

Finally, in addition to qualifying equipment, establish a schedule for requalification to prevent old problems from reoccurring in the manufacturing operation.

Points To Remember

- The problems surrounding automation-caused ESD damage differ considerably from those typically encountered in the manufacturing process.
    - —Charged-device model failures are more prevalent.
    - —The repetitious nature of automation can systematically damage or partially damage large quantities of devices.
    - —Greater engineering skills are required to solve the problems.
- Many companies supplying manufacturing equipment do not understand ESD problems.
- ESD problems in automation are, as a whole, easier to deal with during the design of manufacturing equipment, rather than after the production of the equipment.
- In some cases, it is virtually impossible to institute adequate safeguards after the equipment is fully developed.
- It is necessary to establish written qualification criteria and a plan for equipment qualification similar to those in this chapter.
- The qualification criteria should be included in purchase orders.
- The need for qualifying handling and manufacturing equipment with thorough tests should never be underestimated.
- Manufacturing equipment, as well as the product, is at risk from ESD.
- Maintenance personnel should follow proper ESD procedures at all times.
- It is a "buyer beware" situation.
- Equipment should be periodically requalified to prevent old problems from reoccurring.

Chapter 14

Payback and Benefits

For the expenses associated with ESD control to be justifiable, there must be an economically sound payback and benefit. The measurement of this payback has been an elusive problem for many companies. However, as reported in earlier chapters, we have had considerable success in showing that a ***well-managed ESD control program introduces dramatic improvements***. Financial indicators alone provide sufficient justification for the program, but when the intangible benefits are added, the value of an ESD program becomes overwhelming.

Implementation Cost Control

Economic gains are maximized when implementation costs are controlled. This is accomplished by selecting realistic solutions based on sound technology and placing a very high priority on a well-managed program.

For instance, even though auditing is a vitally important part of managing an effective program, it could be rather costly. However, using basic statistical sampling techniques, 8000 people can be audited by 1 person. The savings in both labor and in reduced ESD losses are significant due to the improved compliance with specified procedures. Used properly, auditing results also make it possible to leverage limited resources effectively. For example, the auditing data can identify those employees having the greatest need for additional training.

The selection of control techniques has a major impact on implementation costs. For instance, during a factory modernization program, we substituted a conductive floor finish for the installation of conductive floor tile and saved ***$5 million*** in the process. The floor tile had been planned for approximately a million square feet, and by substituting the finish, we realized an average cost improvement of $5 per square foot. We have also found through research and experience that dissipative materials provide ample protection in most cases and that the more expensive ***shielding materials are generally unnecessary***. This is another area where substantial sums of money can be saved. However, you do need to know which products are ultrasensitive so that shielding materials can be selectively applied. If your product design and development is well managed, this information will be readily available.

Quality Improvements

Before instituting ESD controls in 1982, AT&T manufacturing suffered losses estimated to be in excess of ***$150 million per year***. The estimate was based on the experience of a number of locations and on the results of a battery of carefully controlled experiments conducted on the manufacturing floor.

Defect levels were known to vary dramatically from one product line to another. In some instances, virtually every device failed, while in others, 2 percent dropouts were recorded. However, all of the product lines, even the more robust ones, were experiencing losses that justified using ESD control measures.

Experimentally, ESD defect levels were found to be ***50 percent of the electrical failures*** (Figure 14-1A). Failure analysis confirmed ESD as the cause of these failures, and after implementing ESD controls, we experienced corresponding yield improvements. In fact, the dramatic improvements were known to greatly enhance productivity and to produce additional test set capacity.

After implementing the systematic controls described in this book, the dramatic quality improvements that had been predicted experimentally became a reality. Locations such as the Denver Works reported in 1980 an annual savings of ***$5 million per year*** with a rate of return on investment of ***3000 percent***. The Reading Works experienced a ***15 percent overall yield improvement***, and Merrimack Valley realized ***$9***

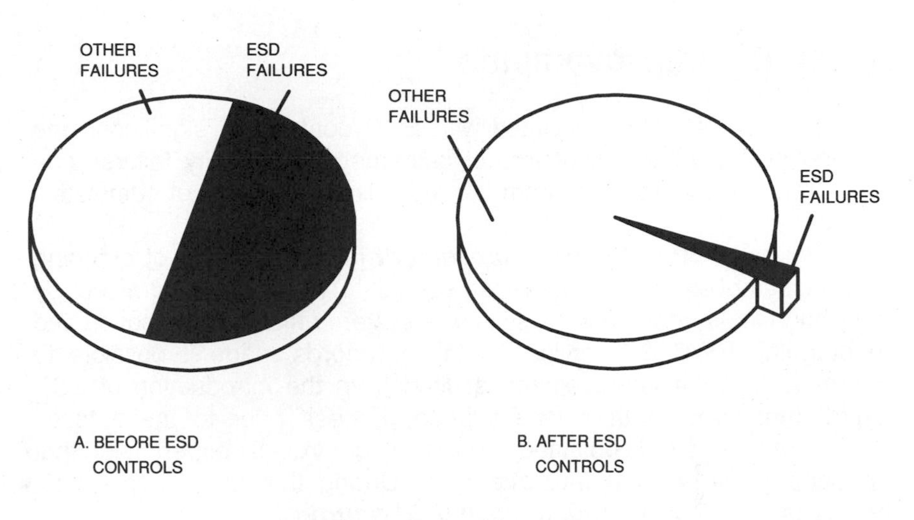

Figure 14-1. AT&T product defect levels

million in savings per year with a rate of return of ***1000 percent***. These savings do not include overhead and therefore reflect the true cost improvements.

Currently, ESD losses at AT&T are minimal. In fact, a recent (1989) review of failures by the Quality organizations (Figure 14-1B) revealed that only three out of three million devices tested failed due to ESD, or ***one in a million***.

The three devices that did fail were analyzed in great detail, and assignable causes were determined. In all three instances, the ***failures were machine-caused***, and none were the result of human error. In one instance, the failure was traced to a charged cable assembly used during the Quality Assurance test procedure. The other two were damaged by the mass soldering machine and specifically by the nylon brushes that are used to clean the PWB assemblies after soldering.

Similar success has been noted at other locations. For instance, the Denver Works reported (in 1989) that ***only 0.23 percent*** of the production failures returned for analysis were ESD induced. The Allentown Works did an analysis of all failures detected during a reliability monitoring operation in 1988 and found ESD to be ***only 1 percent*** of the defects.

Reliability Improvements

The reliability benefits associated with ESD control are significant and are one of the primary concerns of customers. Reliability failures can result from either latency or from the more likely scenario of cumulative damage.

We found a ***three-to-one improvement*** in the reliability of outgoing PWB assemblies over a two-year period. This was determined by analyzing Quality Assurance results of a system reliability test conducted on outgoing PWB assemblies. Detailed records made it possible to determine that the improvement resulted from the introduction of ESD control measures in the manufacturing process prior to these tests. Similar results were obtained during a controlled experiment that produced a ***five-to-one improvement***. During the same time frame, there was a ***25 percent reduction in field returns***.

Intangible Benefits

The financial indicators are impressive, but they do not even begin to compare to the intangible benefits provided by a sound ESD control program. These intangibles stem primarily from ***customer satisfaction*** and the resulting improvement in sales and customer relations.

To protect their investment, informed customers place stringent handling and design requirements on vendors. They are aware of the quality and reliability implications. Furthermore, the handling procedures are a visible sign of a company's commitment to quality. It is therefore, doubly important to meet or exceed the ESD requirements in the interest of customer satisfaction and improved quality.

During visits to the factory, customers often comment on the consistent use of wrist straps and conductive footwear. It has a very positive effect on their perception of product offerings, and for good reason. Consequently, we encourage these observations by challenging them to see if they can find an employee who is not properly grounded — to date no one has. This often leads to customers questioning their own ESD control program and requesting assistance with certain procedures.

This interaction can improve customer relations, especially if your company is a recognized authority in the industry. In regard to ESD control techniques, it is possible to work with customers from the usual perspective of end users on problems in common. In this regard, the experience gained in a manufacturing environment is an invaluable resource because of the unbiased quantitative information readily available. This experience makes it possible to help customers develop their own program and realize the same economic gains while minimizing control costs.

These discussions and interactions are mutually beneficial for customers and suppliers. Customers gain valuable experience in cost-effective controls, improve their service position, and reduce the number of board failures and the associated costs. Suppliers experience an improvement in warranty costs due to fewer field returns. The result is improved customer relations and ultimately increased sales.

Improvement in sales can also be realized by ***differentiating product offerings*** based on ESD handling controls and the designed-in protection described in Chapter 5. In fact, at AT&T both of these issues are formally reviewed during customer audits of the manufacturing process and are part of the supplier evaluation. A negative report in this area has been known to result in lost contracts. On the other hand, a positive review can contribute to the differentiation necessary to secure a substantial contract. Recently, an executive representing a major customer made the following statement, "Whoever comes up with the most robust ESD design will have a competitive edge in selling their products!" Following a plant tour on a separate occasion, another executive said, "The consistent use of wrist straps was impressive, and it is evident that you have a strong commitment to quality. This will have a bearing on our product selection."

Points To Remember

- Maximum savings are gained by an ESD control program when implementation costs are controlled.

- ESD control saves AT&T manufacturing approximately $150 million per year.

- Individual AT&T locations saved up to $9 million per year and realized a 3000 percent return on investment due to ESD control measures.

- The current failure rate at AT&T due to ESD is one part per million with exceptions.

- Machine-caused failures and ultrasensitive devices can still be a significant problem with high failure rates.
- Intangible benefits from an ESD control program are even greater than the tangible gains.
- Wrist straps are a visible sign of the commitment to quality.
- When ESD is not controlled, customer satisfaction is at stake.
- ESD is a source of product differentiation.

Appendix 1

AT&T ESD Policy

Policy

One of the strengths of AT&T products and services has always been its continuing commitment to high quality and reliability. Central to that endeavor is Electrostatic Discharge (ESD) control and prevention. It is, therefore, our policy to:

- Consistently provide components and systems that are sufficiently ESD design hardened to meet the quality and reliability expectations of our customers.
- Consistently adhere to the proper ESD handling, storage, packaging and transportation techniques throughout all facets of our business.

Intent

ESD is known to affect electronic components and systems in a variety of ways and is heavily influenced by technology, design and handling techniques. Therefore, we will strive for continual improvement in prevention techniques, employee awareness, and designed-in-protection at all levels of our company. It is our intention to maintain AT&T as a "World Class" leader in ESD control and design.

This will require the adherence to appropriate design standards as well as controlling the environment around which electronic products are designed, manufactured, transported and used. In order to satisfy this intent we will need to comply with documents such as the AT&T ESD Control Handbook and all appropriate AT&T Product Information Practices. We will strive to provide products that are sufficiently ESD robust in design and free of any latent ESD handling defects to give our customers significant economic advantages in their market.

Responsibilities

Each business group president, entity head, and senior staff officer is responsible for:

- Communicating the policy and seeing that it is carried out effectively.
- Providing adequate support and resources.

Each member of management is responsible for:

- Ensuring adequate design qualification of our product offerings.
- Setting the example by complying with the handling procedures in the "AT&T ESD Control Handbook" at all times.
- Communicating our ESD policy to each employee, visitor and supplier.
- Clarifying specific responsibilities for ESD prevention, awareness and design.
- Establishing effective ESD process controls.
- Ensuring consistent compliance with the AT&T ESD policy and initiating corrective action if needed.

- Implementing and reviewing specific ESD improvement programs.
- Providing education and training in ESD awareness and prevention for all employees.

Appendix 2

Industry, Military, and AT&T Standards Used to Evaluate ESD Control Materials, Equipment, and Devices

Following is a list of the most common standards used to test and evaluate materials, equipment, and devices used in the electronics industry for Electrostatic Discharge testing and control.

AMERICAN ASSOCIATION OF TEXTILE CHEMISTS AND COLORISTS (AATCC)

AATCC-134 — Electrostatic Propensity of Carpets

AMERICAN SOCIETY FOR TESTING AND MATERIALS (ASTM)

ASTM D257-78 — DC Resistance or Conductivity of Insulating Materials

ASTM D790-86 — Flexural Properties of Unreinforced and Reinforced Plastics and Electrical Materials

ASTM D882-82 — Tensile Properties of Thin Plastic Materials

ASTM D2863/88-029 — Measuring the Minimum Oxygen Concentration to Support Candle-Like Combustion of Plastics

ASTM E595-84 — Total Mass Loss and Collected Volatile Condensible Materials from Outgassing in a Vacuum Environment

AMERICAN TELEPHONE AND TELEGRAPH (AT&T AND AT&T BELL LABS)

AT&T Electrostatic Discharge Control Handbook - Issue 2, 1988

PUBS 5100 — Network Equipment Building System (NEBS) Generic Equipment Requirements

T1Y1/88-029 — American National Standard Central Office Equipment—Electrostatic Discharge Requirements

BELL COMMUNICATIONS RESEARCH (BELLCORE)

TA TSY 000870 — Electrostatic Discharge Control in the Manufacturer of Telecommunications Equipment

TR TSY 00078 — Generic Physical Design Requirements for Telecommunications Products and Equipment

TR TSY 000357 — Component Reliability Assurance Requirements for Telecommunications Equipment

ELECTRONICS INDUSTRIES ASSOCIATION (EIA)

EIA-541 — Packaging Material Standards for ESD Sensitive Materials

JEDEC 108 — Distributor Requirements for Handling Electrostatic Discharge Sensitive (ESDS) Devices

ELECTRICAL OVERSTRESS/ELECTROSTATIC DISCHARGE ASSOCIATION (EOS/ESD ASSN.)

EOS/ESD Glossary of Terms

EOS/ESD Draft Standard #1.0 — Personnel Grounding Wrist Straps

EOS/ESD Draft Standard #2.0 — Personnel Garments (Preliminary)

EOS/ESD Draft Standard #3.0 — Ionization (Preliminary)

EOS/ESD Standard #4.0 — Worksurfaces

EOS/ESD Draft Standard #5.0 — Human Body Model (HBM) Test Method (Preliminary)

EOS/ESD Draft Standard #6.0 — Grounding (Preliminary)

INTERNATIONAL ELECTROTECHNICAL COMMISSION (IEC) — (EUROPEAN)

IEC 801-2 — Electromagnetic Compatibility for Industrial Process Measurement and Control Equipment, Part 2: Electrostatic Discharge (ESD) Requirements

NATIONAL ELECTRICAL MANUFACTURERS ASSOCIATION (NEMA)

NEMA #LD-3--85-3.01 — High Pressure Decorative Laminates

NEMA #LD-3.1-85 — Performance, Application, Fabrication, and Installation of High Pressure Decorative Laminates

NATIONAL FIRE PROTECTION ASSOCIATION (NFPA)

NFPA 5-8 — Static Electricity

NFPA 56A — Superceded by NFPA-99

NFPA 77 — Static Electricity

NFPA 99 — Chapters 6 and 7—Standards for Health Care Facilities

UNITED STATES GOVERNMENT (DDO - FED - MIL)

DOD HDBK 263 — Electrostatic Discharge Control Handbook for Protection of Electrical and Electronic Parts, Assemblies and Equipment

DOD-STD-1686 — Electrostatic Discharge Control Program for Protection of Electrical and Electronic Parts, Assemblies and Equipment

DOD-STD-2000-1B — Soldering and Technology, High Quality/High Reliability

DOD-STD-2000-2A — Part and Component Mounting for High Quality/High Reliability Soldered Electrical and Electronic Assemblies

DOD-STD-2000-3A — Criteria for High Quality/High Reliability Soldering Technology

DOD-STD-2000-4A — General Purpose Soldering Requirements for Electrical and Electronic Equipment

FED TEST METHOD STD 101 — Method 4046 - Electrostatic Properties of Materials

MIL-STD-129 — Marking for Shipment and Storage ESD Standard List

MIL-STD-454 — Standard General Requirements for Electronic Equipment (Requirement 5 replaced by DOD-STD-2000)

MIL-STD-701 — List of Standard Semiconductor Devices

MIL-STD-785 — Reliability Program for Systems and Equipment Development and Production

MIL-STD-883 — Method 3015-4 — Electrostatic Discharge Sensitivity Classification

MIL-STD-1285 — Marking of Electrical and Electronic Parts

MIL-STD-1686A — Electrostatic Discharge Control Program for Protection of Electrical and Electronics Parts, Assemblies and Equipment

MIL-E-17555 — Electronic and Electrical Equipment, Accessories, and Provisioned Items (Repair Parts: Packaging of)

MIL-M-38510 — Microcircuits, General Specification for

MIL-STD-45743 E (replaced by DOD-STD-2000)

MIL-D-81705 — Barrier Materials, Flexible, Electrostatic Free, Heat Sealable

MIL-P-81997 — Pouches, Cushioned, Flexible, Electrostatic Free, Reclosable, Transparent

MIL-P-82646 — Plastic Film, Conductive, Heat Sealable, Flexible

MIL-P-82647 — Bags, Conductive Plastic, Heat Sealable, Flexible

PPP-C-1842 — Cushioning Materials, Plastic, Open Cell (for packaging applications)

WS 6536 — (replaced by DOD-STD-2000)

Appendix 3

How to Test the Integrity of the Ground Connection on a Work Surface

Once you have qualified your work surface materials according to industry standards, follow the suggestions below to simplify testing techniques for accepting and installing the material and its ongoing use.

To simplify testing the integrity of the connection between the ground and the inner conductive layer of multilayer work surfaces, we install a second ground bolt (Figure 1). The test is now easy to do with simple equipment. Almost any employee in the manufacturing operation should be capable of using and reading an inexpensive tester such as a wrist strap checker, a continuity checker, or a hand-held, battery-powered multimeter. However, without the second bolt, a more precise and expensive meter is needed to measure the flow of current through the high resistance (10^8 to 10^{10} ohms) dissipative work surface.

The second bolt (test bolt) is installed at a standardized distance (which can be anywhere between two inches and eight inches) from the ground bolt, and then a resistance measurement is made between the bolts with a hand-held multimeter. With the distance fixed, we can predict the reading along with a given tolerance, between the two bolts and through the conductive buried layer (10^3 to 10^5 $\Omega / \square$). Typically, the resistance between bolts eight inches apart will measure less than 500 kohms.

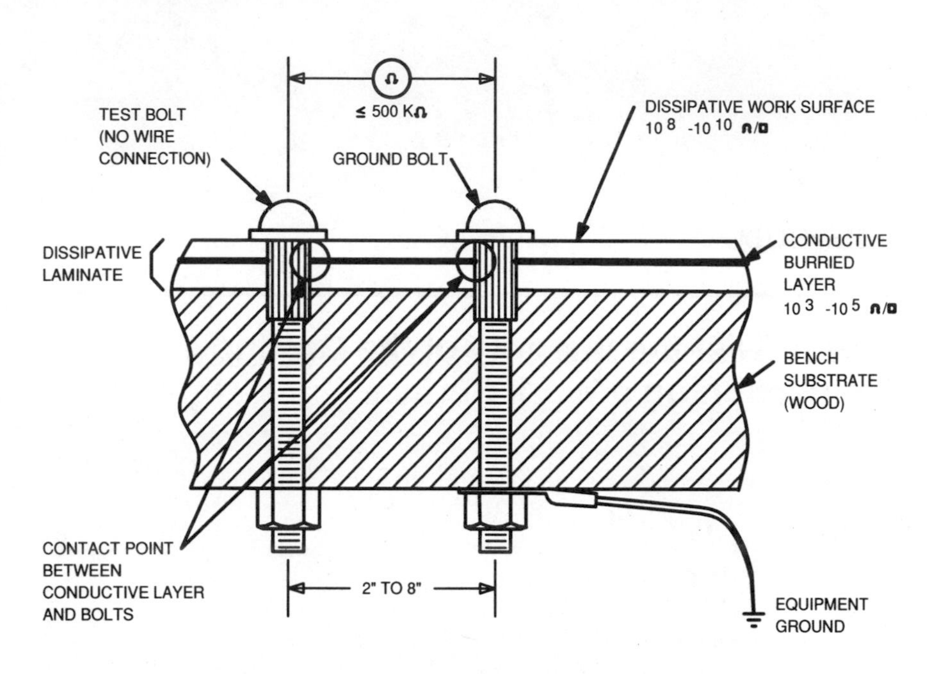

Figure 1. Laminated bench top grounding configuration

Once the work surface has been qualified, we can trust the results of this pass-fail test to check on an on-going basis for a gross failure with the integrity of the connection between the conductive layer and the ground bolt. A wrist strap checker will detect if the resistance is below approximately 750 kohms. Since the conductive layer should be less than 500 kohms, a red light signals an acceptable ground connection between the ground bolt and the conductive buried layer. A continuity test from the ground bolt to equipment ground will, now, verify that the work surface is properly grounded..

Glossary

The following terms are used in this book.*

Air Ionizers: A source of charged air molecules (ions). The positive ions are attracted to negatively charged bodies and the negative ions to positively charged bodies, resulting in charge neutralization. (See also ***Ionization***.)

Ampere: The unit of electrical current.

Antistat, Antistatic Agent: A chemical compound that, when impregnated in or topically applied to a primary material or substrate, renders the primary material ***antistatic***.

Antistatic: A property of materials that resist ***triboelectric charging*** and produce minimal ***static charge*** when separated from themselves or other materials. A material's antistatic property is not necessarily correlated with its ***resistivity***.

Antistatic Plastic: Any specialized plastic that reduces ***triboelectric charging***. Although some ***antistat agents***, especially hygroscopic ones, might slightly increase the surface conductivity of the base plastic, the resultant conductivity is usually insufficient to provide ***Faraday Cage shielding***. (See also ***Topical Antistats***.)

Antistatic Property: The characteristic of an item that allows the item to effectively minimize the production of a static charge when materials are separated from another surface. This characteristic is not a dependent function of material resistivity or static decay performance.

---

* For a more complete definition of terms used in the industry, refer to the EOS/ESD Glossary of Terms.

Astatic (European terminology): A distinction in resistivity level. In terms of increasing resistivity: antistatic, astatic, static. These terms have been used over the years in European literature.

Avalanche Breakdown: A breakdown caused by the cumulative multiplication of charge carriers through field-induced impact ionization.

Bare PWB: A printed wiring board that ***has not*** been populated with components.

Bipolar Transistor: A device containing junction semiconductors that uses both positive (p-type) and negative (n-type) charge carriers.

Catastrophic Failure: A failure that is both sudden and complete and that involves complete loss of the required function. Catastrophic ***ESD*** failures are the result of ***electrical overstress (EOS)*** caused by an ***electrostatic discharge***.

CDM: Charged-device model.

Charge: An excess or deficiency of electrons on the surface of material, measured in units of coulombs.

Charged-Device Model: A model characterizing an ***ESD*** in which a device isolated from ground is charged and then suddenly discharged.

CMOS: An integrated circuit that uses complementary n-channel and p-channel metal oxide semiconductor transistors.

Component: A semiconductor device.

Conductive: A property of materials which are either metal or impregnated with metal, carbon particles or other conductive materials, or whose surface has been treated with such materials through a process of lacquering, plating, metalizing, or printing. A ***conductive material*** for static control purposes shall have a ***surface resistivity*** less than 1×10^5 ohms/square or 1×10^3 ohm-cm if volume conductive. A ***conductive material*** is not necessarily antistatic.

Conductor: A material that allows a current of electrons to continuously pass along it or through it when a voltage is applied. These materials exhibit low resistance.

Coulomb: The unit or quantity of electricity or charge. One coulomb is the quantity of charge transferred by a current of 1 ampere in 1 second.

CPM: Charge plate monitor.

Cumulative Failure: A device failure resulting from multiple exposures to ESD.

DC: Direct current.

Decay Time: The time between two specified values of a variable that decreases with time.

Device: A package of electronic circuitry. This term is often interchangeably used with "semiconductor" or "component" when describing units sensitive to ESD damage.

Dielectric Breakdown: A threshold effect in a dielectric medium where at some electric field strength across the medium, bound electrons become unbounded and travel through the medium as a current. In solid media, the region of the current path can be permanently damaged. The units of measurement are usually volts per unit of thickness.

DIP (Dual In-line Package): A type of housing for integrated circuits consisting of a molded plastic or ceramic container with two rows of pins.

Dissipative: Material exhibiting a surface resistance of 10^5 through 10^{12} ohms per square. Dissipative materials bleed off charges at an optimal rate, neither too fast nor too slow.

DOA (Dead-On-Arrival): A failure of a device that occurs in the device's initial usage.

DUT: Device under test.

ECL: Emitter-coupled-logic microcircuit.

Electrical Overstress (EOS): The electrical stressing of items beyond their specifications. May be due to ***ESD***.

Electromagnetic Compatibility (EMC): The capability of electronic equipment or systems to function in the intended operational electromagnetic environment.

Electromagnetic Interference (EMI): Impairment of a desired electromagnetic signal by an electromagnetic disturbance.

Electromagnetic Pulse (EMP): A pulse generated by transient electromagnetic fields.

Electromagnetic Shield: A screen or other housing placed around devices or circuits to reduce the effects of external electric and magnetic fields.

Electron: A negatively charged particle with an electrical charge equal to approximately 1.6×10^{19} coulomb.

Electrostatic Charge: See ***Static Electricity***.

Electrostatic Damage: Damage to a device from ***static electricity*** by either (1) ***discharge*** from a charged conductor, (2) electric field induction, or (3) self discharge.

Electrostatic Discharge (ESD): The transfer of ***electrostatic charge*** between bodies at different ***electrostatic potentials***.

Electrostatic Field: The region surrounding an electrically charged object in which another electrical charge will experience a force. Quantitatively, it is the voltage gradient between two points at different potentials.

Electrostatic Potential: The potential difference between a point and an agreed-upon reference.

Electrostatic Shield: A barrier or enclosure that prevents the penetration of an electrostatic field. An ***electrostatic shield*** may not offer much protection against the effects of electromagnetic fields.

Electrostatic: The study of properties and behavior of electrical charge at rest.

EIA: Electronic Industries Association.

EMI: Electromagnetic interference.

EOS: Electrical overstress.

EPROM: Erasable programmable read-only memory.

EPS: Expanded polystyrene.

ESD (Electrostatic Discharge): A sudden transfer of charge between two objects.

ESD Susceptibility: A measure of the susceptibility of an item to ***ESD damage***. (See also ***ESDS Device Classification***.)

ESDS (Electrostatic Discharge-Sensitive): Describes devices that are vulnerable to damage from electrostatic discharge.

ESDS Device Classification: ESDS items are divided into classes based on their specific test failure voltages.

ESDS Item: Electrical and electronic parts, assemblies, and equipment that are sensitive to ***ESD voltage***.

EUT: Equipment under test.

Failure Mechanism: The process by which a device is caused to fail.

Failure Mode: The effect by which a failure is observed; for example, an open circuit.

Faraday Cage: An electrically continuous conductive enclosure that provides ***electrostatic shielding***.

FCDM: Field-induced, charged-device model.

Field-Induced Model: A model used to test for ***ESDS*** that simulates a situation where an electrically floating device is subjected to an ***electrostatic field*** and then contacted to a conductive object.

Film Resistor: A resistor made by deposition of a resistive metal or compound onto a substrate.

Floating Conductor: A conductor that is isolated from the grounding system within a static controlled environment.

FMA: Failure mode analysis.

GaAsFET: Gallium-arsenide field effect transistor.

Ground: A metallic connection with the earth to establish zero potential or voltage with respect to ground or earth. It is the voltage reference point in a circuit. There may or may not be an actual connection to earth, but it is understood that a point in the circuit said to be a ***ground*** potential could be connected to earth without disturbing the operation of the circuit.

Grounding: Connecting to a ground or to a conductor that is grounded.

Ground Straps: A wrist, leg, or ankle strap that discharges ***static charges*** on the human body safely to ground and equalizes personnel static levels with that of the work surface. Straps should be current-limited for safety.

HBM: Human body model.

HIC: Hybrid integrated circuit.

Human Body Model: A circuit that simulates the ***ESD*** from a person for testing purposes.

IEC: International Electrotechnical Commission.

Inductive Charging: See ***Field-Induced Model***.

Input Protection: A protective network at the input pins of an item that attempts to prevent damage from ***ESD***.

Insulator: A material that does not conduct electricity. Also known as "dielectric" material.

Insulation Resistance (Ri): The ratio of the direct voltage applied to two electrodes that are in contact with or embedded in a specimen to the total current between them. It is dependent on both the ***volume and surface resistivity*** of the specimen.

Insulative Material: A material having a surface resistivity of at least 1×10^{12} ohms/square, or 1×10^{10} ohm-cm volume resistivity.

Integrated Circuit (IC): A monolithic electron device containing transistors, resistors, capacitors, etc., in a single package.

Ionization: The process by which a neutral atom or molecule, such as air, acquires a positive or negative charge.

JFET: Junction field effect transistor.

Latent Failure: A hidden failure that did not result in an immediately detectable defect condition.

LDD: Lightly doped drain.

Lid: Metal or ceramic ends or covers for integrated circuit packages.

MOSFET: Metal-oxide-semiconductor field effect transistor.

NEMA: National Electrical Manufacturers Association.

NFPA: National Fire Protection Association.

NMOS: N-type metal oxide semiconductor.

Nonconductor: Usually, a dielectric or insulating material that breaks down and begins to conduct current when a high enough voltage is applied.

Ohm: The unit of electrical resistance. It is the resistance through which a current of 1 ampere flows when a voltage of 1 volt is applied.

Ohms per Square: A unit of surface resistance. The surface resistance of a material is numerically equal to the resistance between two electrodes forming opposite sides of a square. The size of the square is immaterial.

Op Amp: Operational amplifier.

Potential: Stored energy that is able to do work and is measured in millivolts, volts, or kilovolts (kV).

PVT: Personnel voltage tester.

PWB: Printed wiring board.

PWB assembly: A printed wiring board that ***has*** been populated with components.

Radio Frequency Interference (RFI): A form of ***electromagnetic interference (EMI)***. Any electrical signal capable of being propagated and interfering with the proper operation of electrical or electronic equipment. The spark from a static discharge is a source of ***RFI***.

Resistance: The degree of "difficulty" that an electrical current encounters in passing through a circuit or conductor. Resistance is measured in ohms and is a property of conductors that is determined by conductor dimensions, materials, and temperature. The resistance of a material determines the current produced by a given voltage.

RTG: Resistance to ground.

RVM: Residual voltmeter.

Schottky Diode: A metal-semiconductor rectifying contact. It is often used in transistor-transistor-logic (TTL) integrated circuits to shunt the base collector junction of transistors. It forms a clamped composite transistor with a very short saturation time constant.

SCR: Silicon-controlled rectifier.

Semiconductor: Any class of solid (such as germanium or silicon) whose electrical conductivity is similar to that of metals at high temperatures and virtually absent at low temperatures.

SEM: Scanning electrical microscope.

Shop: A manufacturing department or designated work area assigned to one functional manager.

SRM: Surface resistivity meter.

Static Charge: See ***Charge***.

Static Electricity: Electrical charge at rest.

Static Eliminator — Induction: Induction static eliminators generally consist of a series of conductive grounded points or brushes. When a single sharp grounded-needle is brought into the proximity of any high-charged surface, it has induced in it a charge opposite to that of the surface. When a high enough charge concentration has been developed, the surrounding air will break down. A vast number of charge balancing ions are formed. The simple "tinsel string" static eliminator is an example of an induction static eliminator.

Static Eliminator — Nuclear: Nuclear static eliminators create ions by the irradiation of the air molecules. Some models use an alpha particle emitting isotope to create sufficient ion pairs to neutralize a charged molecule. The high speed particle interacts with the air molecules with sufficient energy to actually strip off one of its outer electrons. (See also ***Ionization*** and ***Air Ionizers***.)

Static-Safe: Conditions or materials that provide adequate protection for the product sensitivity involved. The material test specifications are defined in EIA-541, *Packing Material Standards for Protection of Electrostatic Discharge Sensitive Devices,* from the Electronic Industries Association.

Surface Resistivity (ρ_s): The ratio of a DC voltage to the current that passes across the surface of the system. In this case, the surface consists of a square unit of area. In effect, the surface resistivity is the resistance between two opposite sides of a square and is independent of the size of the square or its dimensional units. Surface resistivity is expressed in ohms/square. When using a concentric ring fixture, resistivity is calculated by using the following expression:

$$\textit{Surface Resistivity } (\rho_s) = \left[\frac{\pi(D2+D1)}{(D2-D1)}\right] \times R$$

Where D2 = Inside diameter of outer electrode
D1 = Outside diameter of inner electrode
R = Measured resistance

Surfactant: See ***Antistatic Agent*** and ***Topical Antistats***.

Topical Antistats: Chemical agents that when applied to surfaces of ***insulative materials*** reduces their ability to generate static.

Triboelectric: A term referring to a static charge generated by friction or the separation of materials.

Triboelectric Charging: The generation of ***electrostatic charges*** when two pieces of material in intimate contact are separated (where one or both is an insulator). Substantial generation of ***static electricity*** can be caused by contact and separation of two materials or by rubbing two substances together. (See also ***Triboelectric Series.***)

Triboelectric Series: A list of substances arranged so that one can become positively charged when separated from one farther down the list, or negatively charged when separated from one farther up the list. The series is used primarily to indicate likely resultant charge polarities after triboelectric generation. However, this series is derived from specially prepared and cleaned materials tested in very controlled conditions. In everyday circumstances, materials reasonably close to one another in the series can produce charge polarities opposite to that expected. This series is only a guide.

VCP: Vertical conducting plane.

Volt: The unit of electromotive force or potential. One volt sends a current of 1 ampere through a resistance of 1 ohm.

Voltage Suppression: A phenomenon where the voltage from a charged object is reduced by increasing the capacitance of the object rather than decreasing the charge on the object. The relation Q=CV describes the phenomenon.

Volume Resistivity (ρ_v): The ratio of the DC voltage per unit thickness applied across two electrodes in contact with a specimen to the amount of current per unit area passing through the system. Volume resistivity is generally given in ohm-centimeters. When using a concentric ring fixture, resistivity is calculated by using the following expression:

$$\textit{Volume Resistivity}\ (\rho_v) = \left[\frac{\pi D1^2}{4T}\right] \times R$$

Where D1 = Diameter of inner electrode or disc
R = Measured resistance in ohms
T = Thickness of specimen.

VMOS: Vertical metal-oxide semiconductor.

VSLI: Very large scale integration.

References

1. Gagne, Robert, *Essentials of Learning for Instruction,* Prentice Hall, 1988.
2. McAteer, O. J., "ESD - A Decade of Progress," *EOS/ESD Symposium Proceedings,* EOS-10 (1988).
3. McFarland, W. Y., "The Economic Benefits of an Effective ESD Awareness and Control Program - An Empirical Analysis", *EOS/ESD Symposium Proceedings,* EOS-3 (1981).
4. Euker, R., "ESD in I. C. Assembly (A Baseline Solution)", *EOS/ESD Symposium Proceedings,* EOS-4 (1982), p. 142.
5. Dangelmayer, G. T., "ESD - How Often Does It Happen?", *EOS/ESD Symposium Proceedings,* EOS-5 (1983), p. 1.
6. Downing, M. H., "Control Implementation and Cost Avoidance Analysis", *EOS/ESD Symposium Proceedings,* EOS-5 (1983), p. 6.
7. Lindholm, A., "A Case History of an ESD Problem", *EOS/ESD Symposium Proceedings,* EOS-7 (1985), p. 10.
8. Halperin, S. A., "Estimating ESD Losses in the Complex Organization", *EOS/ESD Symposium Proceedings,* EOS-8 (1986), p. 1.
9. Frank, D. E., "The Perfect '10' - Can You Really Have One?", *EOS/ESD Symposium Proceedings,* EOS-3 (1981), p. 21.
10. Strand, C. J., A. Tweet and M. E. Weight, "An Effective Electrostatic Discharge Protection Program", *EOS/ESD Symposium Proceedings,* EOS-4 (1982), p. 145.
11. Kirk, W. J., "Uniform ESD Protection in a Large Multi-department Assembly Plant", *EOS/ESD Symposium Proceedings,* EOS-4 (1982), p. 165.

12. Dangelmayer, G. T., "A Realistic and Systematic ESD Control Plan", *EOS/ESD Symposium Proceedings,* EOS-6 (1984), p. 1.

13. Lai, E. and J. Plaster, "ESD Control in the Automotive Electronics Industry - A Case Study", *EOS/ESD Symposium Proceedings,* EOS-9 (1987), p. 1.

14. McAteer, O. J., "An Effective ESD Awareness Training Program", *EOS/ESD Symposium Proceedings,* EOS-1 (1979), p. 1.

15. Schnetker, T. R. "Human Factors in Electrostatic Discharge Protection", *EOS/ESD Symposium Proceedings,* EOS-1 (1979), p. 122.

16. Halperin, S. A., "Facility Evaluation: Isolating Environmental ESD Problems", *EOS/ESD Symposium Proceedings,* EOS-2 (1980), p. 192.

17. McAteer, R. E., G. H. Lucas and A. McDonald, "A Pragmatic Approach to ESD Problem Solving in the Manufacturing Environment, A Case History", *EOS/ESD Symposium Proceedings,* EOS-3 (1981), p. 34.

18. Hansel, G. E., "The Production Operator: Weak Link or Warrior in the ESD Battle?", *EOS/ESD Symposium Proceedings,* EOS-5 (1983), p. 12.

19. Dangelmayer, G. T. and E. S. Jesby, "Employee Training for Successful ESD Control", *EOS/ESD Symposium Proceedings,* EOS-7 (1985), p. 20.

20. R. J. Zezulka, "Tracking Results of an ESD Control Program Within a Telecommunications Service Company", *EOS/ESD Symposium Proceedings,* EOS-11, 1989, p. 36.

21. Kirk, W. J. Jr., L. S. Carter and M. Waddell, "Eliminate Static Damage to Circuits", *Electronic Design,* March 29, 1976.

22. "An ESD Management Focus," *EOS/ESD Association,* 1989.

23. Halliday, D., and R. Resnick, *Fundamentals of Physics,* Second Edition, Wiley, New York, 1981.

24. Hayt, W. H., *Engineering Electromagnetics,* Fourth Edition, McGraw-Hill, New York, 1981.

25. Moore, A. D., *Electrostatics,* Doubleday, Garden City, NY, 1968.

26. Ku, C. C., and R. Liepins, *Electrical Properties of Polymers,* Hanser Publishers, Munich, 1987.

27. Dangelmayer, G. T. "ESD—How Often Does It Happen?" *EOS/ESD Symposium Proceedings,* EOS-5, 1983, p. 1.

28. Jon, M. C., D. Robinson-Hahn, and T. L. Welsher, "Tape and Reel Packaging—An ESD Concern," *EOS/ESD Symposium Proceedings,* EOS-10, 1988, p. 15.

29. Bossard, R. P., R. G. Chemelli, and B. A. Unger, "ESD Damage from Triboelectrically Charged Pins," *EOS/ESD Symposium Proceedings,* EOS-2, 1980, p. 17.

30. Avery, L. R., "Charged Device Model Testing: Trying to Duplicate Reality," *EOS/ESD Symposium Proceedings,* EOS-9, 1987, p. 88.

31. Lafferty, D., "Secondary Discharge: A New Jeopardy and New Tool," *EOS/ESD Symposium Proceedings,* EOS-6, 1984, p. 131.

32. Maloney, T. J., "Designing MOS Inputs and Outputs to Avoid Oxide Failure in the Charged-Device Model," *EOS/ESD Symposium Proceedings,* EOS-10, 1988, p. 220.

33. Chemelli, R. G., B. A. Unger, and P. R. Bossard, "ESD by Static Induction," *EOS/ESD Symposium Proceedings,* EOS-5, 1983, p. 29.

34. Speakman, T. S., "A Model for the Failure of Bipolar Silicon Integrated Circuits Subjected to Electrostatic Discharge," *12th Annual Proceedings Reliability Physics,* April 1974, p. 60.

35. Unger, B. A., *"Electrostatic Discharge (ESD) Failure Mechanisms and Model",* Nepcon West, Anaheim, California, February 23-25, 1982.

36. McAteer, O. J., and R. E. Twist, "Latent ESD Failures," *EOS/ESD Symposium Proceedings,* EOS-4, 1982, p. 41.

37. Antinone, R. J., P. A. Young, D. D. Wilson, W. E. Echols, M. G. Rossi, W. J. Orvis, G. H. Khanaka and J. H. Lee, *Electrical Overstress Protection for Electronic Devices,* Noyes Publications, Park Ridge, New Jersey, 1986.

38. Pelella, A. R., "Design of ESD Protection Networks for Integrated Circuits," Doctoral Dissertation, Clarkson University, Potsdam, NY, 1985.

39. Minear, R. L., and G. A. Dodson, "The Phantom Emitter—an ESD-Resistant Bipolar Transistor Design and its Applications to Linear Integrated Circuits," *EOS/ESD Symposium Proceedings,* EOS-1, 1979, p. 188.

40. Lin, C. M., "A CMOS VLSI ESD Input Protection, DIFIDW," *EOS/ESD Symposium Proceedings,* EOS-6, 1984, p. 202.

41. Duvvury, C., R. N. Rountree, Y. Fong, and R. A. McPhee, "ESD Phenomena and Protection Issues in CMOS Output Buffers," *25th Annual Proceedings Reliability Physics,* 1987, p. 174.

42. Polgreen, T., and A. Chatterjee, "Improving the ESD Failure Threshold of Silicided NMOS Output Transistors by Ensuring Uniform Current Flow," *EOS/ESD Symposium Proceedings,* EOS-11, 1989, p. 167.

43. Rieck, G., and R. Manely, "Novel ESD Protection for Advanced CMOS Output Drivers," *EOS/ESD Symposium Proceedings,* EOS-11, 1989, p. 182.

44. Ott, H. W., *Noise Reduction Techniques in Electronic Systems,* Second Edition, Wiley-Interscience, 1988.

45. Mardiguian, M., *Electrostatic Discharge,* Interference Control Technologies, Inc., Gainesville, VA, 1986.

46. Boxleitner, W., *Electrostatic Discharge and Electronic Equipment,* IEEE Press, New York, 1988.

47. "Electrostatic Discharge Sensitivity Testing," *MIL-STD 883C,* Method 3015.6, February 1989.

48. "Human Body Model (HBM) Electrostatic Discharge Sensitivity Testing," EOS/ESD-DS-5.0, Preliminary Draft Standard, EOS/ESD Association, Rome, NY, 1989.

49. Chemelli, R. G., and L. F. DeChiaro, "The Characterization and Control of Leading Edge Transients from Human Body Model ESD Simulators," *EOS/ESD Symposium Proceedings,* EOS-7, 1985, p. 155.

50. Lin, D. L., M. S. Strauss, and T. L. Welsher, "On the Validity of ESD Threshold Data Obtained Using Commercial Human-Body Model Simulators," *25th Annual Proceedings Reliability Physics,* 1987, p. 77.

51. Duvvury, C., R. N. Rountree, and O. Adams, "Internal Chip ESD Phenomena Beyond the Protection Circuit," *IEEE Trans. Electron,* Dev. 35, 1988, p. 2133.

52. Renninger, R. G., D. L. Lin, M. C. Jon, T. Diep, and T. L. Welsher, "A Field-Induced Charged-Device Model Simulator," *EOS/ESD Symposium Proceedings,* EOS-11, 1989, p. 59.

53. Anderson, D. C., "Conductor Contact, a cause of ESD Damage," *Appliance Engineering,* Vol 7, p. 49 (1990)

54. Huntsman, J. R., "Triboelectric Charge: Its ESD Ability and a Measurement Method for its Propensity on Packaging Materials", *EOS/ESD Symposium Proceedings,* EOS-6, 1984, p. 64.

55. Unger, B. A., R. G. Chemelli, P. R. Bossard, and M. R. Hudock, "Evaluation of Integrated Circuit Shipping Tubes", *EOS/ESD Symposium Proceedings,* EOS-3, 1981, p. 57.

56. Unger, B. A., and D. L. Hart, "Triboelectric Characterization of Packaging Materials", *EOS/ESD Symposium Proceedings,* EOS-7, 1985, p. 107.

57. Jon, M. C., D. Robinson-Hahn, and T. L. Welsher, "Tape and Reel Packaging - An ESD Concern", *EOS/ESD Symposium Proceedings,* EOS-10, 1988, p. 15.

58. Yenni, D. M., and J. R. Huntsman, "The Deficiences in Military Specification MIL-B-81705: Considerations and a Simple Model for Static Protection", *EOS/ESD Symposium Proceedings,* EOS-1, 1979, p. 45.

59. Huntsman, J. R., and D. M. Yenni, "Test Methods for Static Control Products", *EOS/ESD Symposium Proceedings,* EOS-4, 1982, p. 94.

60. Kolyer, J. M., and W. E. Anderson, "Selection of Packaging Materials for Electrostatic Discharge-Sensitive (ESDS) Items", *EOS/ESD Symposium Proceedings,* EOS-3, 1981, p. 75.

61. Holmes, G. C., P. J. Huff, and R. L. Johnson, "An Experimental Study of the ESD Screening Effectiveness of Anti-static Bags", *EOS/ESD Symposium Proceedings,* EOS-6, 1984, p. 78.

62. Chemelli, R. G., B. A. Unger, and P. R. Bossard, "ESD by Static Induction", *EOS/ESD Symposium Proceedings,* EOS-5, 1983, p. 29.

Index

J

K

L

M

N

O

P

S

U

V

W

Y

Z